SALLIE E. WILDY
B: Nov. 3, 1849
P: Yazoo Co., MS
D: Feb. 20, 1884
P: Roswell (Chaves) NM

William W. WILDY
B: Oct. 9, 1809
P: Nrthmbrlnd Co. VA
D: Feb. 7, 1881
P: Yazoo Co., MS

William WILDY
B: 1760
P:
D: By Sept. 1818
P: Northumberland Co., VA

Judith HARDING
B: July 18, 1745
D:

John C. COTTRELL
B: 1660
D: Before Jan. 19, 1727
P: NC or VA?

Elizabeth (?)

Sarah (ADERTON?)
B:
D:

M. Laurena HALL
B: Aug. 3, 1819
P: Nashville, TN
D: Nov. 13, 1871
P: Yazoo Co., MS

Samuel S. HALL
B: c. 1790-1800
D: Before Sept. 1842
P: Yazoo Co., MS

William HALL
B: c. 1763
D: Aug., 31, 1838

Matilda GARRETT
B: 1808
P: Davidson Co., TN
D:

Jacob GARRETT
B: 1776
P: Pr. Wm. Co., VA
M: Mar. 12, 1802
D: July 30, 1842

Charity TAYLOR
M: Mar. 12, 1802
P: Davidson Co., TN
D: 1840
P: San Augustine, TX

John TAYLOR

Jane (?)

CAPTAIN JOSEPH C. LEA

From Confederate Guerrilla

To New Mexico Patriarch

Captain Joseph C. Lea, *c.* 1887
Photo by Rhine, Fort Worth, Texas
Courtesy: HCSNM, Photo No. 356-C

CAPTAIN JOSEPH C. LEA

From Confederate Guerrilla To New Mexico Patriarch

Elvis E. Fleming

Yucca Tree Press
Las Cruces, New Mexico
In cooperation with
Historical Society for Southeast New Mexico

 Yucca Tree Press, 2130 Hixon Drive, Las Cruces, New Mexico, 88005-3305.

First Printing - March 2002

Fleming, Elvis E.
CAPTAIN JOSEPH C. LEA: From Confederate Guerrilla to New Mexico Patriarch
1. Joseph C. Lea, 1841-1904. 2. New Mexico - Territorial History. 3. United States History - Civil War, 1861-1865. 4. Roswell, New Mexico - History
I. Elvis E. Fleming. II Title

Library of Congress Control Number: 2001099356
ISBN: 1-881325-53-9

FOREWORD

Having dabbled in the history of Southeast New Mexico for the past twenty-five years, I often observed that one individual appeared again and again as the primary force for change: Joseph C. Lea. Others—such as J.J. Hagerman, Charles B. Eddy, Pat Garrett, John Poe, George Curry, Nathan Jaffa—played undeniably key roles; but none approached Lea in terms of sustained involvement in virtually every positive change—economic, political, educational—in the region.

Because of these observations, I've frequently supported local historian Elvis E. Fleming's observation that we need a biography of this force most pervasive not only in Roswell and Southeast New Mexico but in the Territory itself. Those of us who strongly believe that identification with a community and support for a region are tied to a sound familiarity with the area's history were delighted when Fleming, recently retired, decided to take up the challenge of a Lea biography.

Fleming consulted virtually every relevant primary document including materials from numerous archival collections and seemingly every piece of published literature even remotely pertaining to Lea in compiling this account. Certainly, the times in Lea's youth including his involvement in the Confederate cause in the Civil War with Quantrill and Shelby's guerrilla operations in the West helped shape the young man. Fleming's work suggests, however, that with Lea's arrival at the confluence of the Hondo and Pecos in 1877, this man shaped his times.

Few, if any, Lea-like movers and shakers work in isolation, and Fleming develops the physical and social context within which this extraordinary frontier figure effected improvements. While contending with occasional marauding Indians, keeping his distance from the Lincoln County imbroglio, promoting the area in Santa Fe, Lea through commerce, ranching, and real estate transformed Van C. Smith's struggling little hamlet into a formidable turn-of-the-century territorial city. Fleming notes that Lea even tapped into a significant chunk of eastern capital, that of financier Horace Thurber, to underwrite economic growth.

Fleming further explores the political dimension of Lea's impact. Lea's Civil War leadership experiences—after all he was forever known as 'Captain Lea'—and apparent financial successes made him the logical individual to speak for this rapidly growing cluster of people on the Pecos. At six feet and four inches, Lea was not only an imposing presence but also an individual of considerable intelligence and communicative capacity who obviously could convince others to invest in his causes. Today we would probably refer to him as charismatic. This frontier figure also exhibited considerable capacity to identify and to coordinate with political allies. His work with George Curry, later elected governor, provides a good example of just that diplomatic skill. Fleming, through his work on Lea, in effect, presents a case study on how New Mexico politics operated on the territorial, regional and local levels.

Although Fleming doesn't directly mention it, there's little doubt that Lea's extended family of brothers, sisters, numerous in-laws and their offspring constituted a significant support group. Certain members, most significantly his third wife, Mabel Doss Day, directed Lea's numerous talents toward educational improvements, specifically the founding of New Mexico Military Institute. Fleming's research on Lea's immediate family and its subsequent difficulties offer an especially interesting social dimension to this work.

A man's accomplishments might be measured by what still stands nearly a century after his death. With even international recogni-tion, Roswell would likely make its 'Father' quite proud. New Mexico Military Institute continues to enjoy a national reputation as one of the finest schools of its kind, confirming the perseverance of its continued benefactor: Capt. Joseph C. Lea. Elvis E. Fleming has made an invaluable contribution both to compiling an important body of historical information on New Mexico and to providing insight into one of those special individuals who made good things happen.

William E. Gibbs, Ph. D.
New Mexico Military Institute
Roswell, New Mexico
May 2001

ACKNOWLEDGMENTS

My quest for information about Joseph C. Lea and his various activities has taken me down many roads—some on the Information Super-Highway. I am continuously amazed at the information and contacts possible through the Internet, all of which greatly facilitated the writing of this book.

I am indebted to the noted British historian and writer, Frederick Nolan, for suggesting to me that someone should write a book about J. C. Lea. Mr. Nolan is the leading authority on the Lincoln County War, and I am extremely grateful that he took time to evaluate and critique a part of the chapter on Lea's role in that war and its aftermath.

A number of descendants of Lea and of his parents-in-law, Major William W. Wildy and Laurena Hall Childress Wildy, have made invaluable contributions to this work. Captain Lea's granddaughter, Josephine Bedell Rohr, and her daughter, Lea Rohr Fraser, both of Roswell, New Mexico, placed a small collection of papers and photographs of the family that had been collected by Lea's daughter, Ella Lea Bedell Dow, in the archives of the Historical Center for Southeast New Mexico in Roswell. Those papers, together with many other sources in the archives, formed an excellent starting place for my research.

Other family members who made major contributions are Barbara 'Bobbie' Buckner and her daughter, Verna Reed, both of Carlsbad, New Mexico; Lawrence W. Long, III, of Yazoo City, Mississippi, a descendant of Laurena Wildy; Thomas Lea Garry of San Gabriel, California, who is a descendant of Harry Wildy Lea; and Tom Lea, noted artist and writer of El Paso and a descendant of J.C. Lea's brother Thomas.

Historian Donald Hale of Lee's Summit, Missouri, was very help-ful on the Civil War chapter, which he also critiqued for me. Dr. William E. Gibbs of New Mexico Military Institute not only provided information on the school; he also critiqued the NMMI chapter, read the entire manuscript, and agreed to write the Foreword for the book.

Besides the archives of the Historical Center, many other repositories and libraries have furnished valuable assistance. Among those are the National Archives; the state archives of New Mexico, Missouri, Wyoming, and Mississippi; the Jackson County Historical Society Archives in Independence, Missouri; the State Historical Society of Missouri; the offices of the Chaves County District Clerk, County Clerk, and County Tax Assessor in Roswell; the county clerk's offices in Lincoln, Colfax, Taos, and Mora counties; and the Pecos Valley Artesian Conservancy District, Roswell.

Helpful libraries include Eastern New Mexico University-Roswell, Roswell Public Library, New Mexico Military Institute Library, and Denver Public Library. A number of libraries in Missouri have provided helpful information, including those at Lee's Summit, Independence, St. Joseph, and the University of Missouri -Columbia. Unfortunately, the documents formerly in the New Mexico Military Institute archives which pertain to the founding of the school have been lost, so I was not able to examine those sources first hand.

Several other individuals not already mentioned have been generous with their knowledge and resources that have helped me reconstruct the life of J. C. Lea. These include Jim Bullock, Ernestine Chesser Williams, Bob Roen, Peg Stokes, Bernice D'Abadie, Morgan Nelson, Max Littell, and Mary Herrington, all of Roswell. Others were historian Gene M. Gressley of Jacksonville, Oregon; and Pete Eckermann of El Paso.

While I have been able to obtain the most important information from the letters of J.C. Lea, Mabel Day Lea, and H.K. Thurber from other sources, I have not been permitted by the family of James T. Padgitt to examine those papers myself. Padgitt, who was Mabel Doss Day Lea's grandson, whote some articles based on the letters, and Gene M. Gressley used the letters in writing his history of western ranching. Some of the unanswered questions about the relationship between Lea and his third wife and about the final years of the Lea Cattle Company might be answered in those papers, but unfortunately I did not have an opportunity to see them.

I have received support from others, such as Michael Francis, editor of *Vision Magazine;* the board of directors of the Historical Society for South-east New Mexico and the staff of the Historical Center, Sharon Jaskiewicz and Roger Burnett; and Janie Matson of Yucca Tree Press. Finally, my bride of forty-six years, Menza Holden Fleming, has fulfilled the roles of encourager, grammar expert, and critic, for which I am eternally grateful.

I have not attempted to write a 'scholarly' book with footnotes, preferring instead to mention the sources in the text. Further, I have tried to use a style that the non-historian might find readable. It is my desire that the message of this book will help residents of New Mexico develop a greater appreciation for the work of pioneers such as J.C. Lea. The city, all of the Pecos Valley, and New Mexico in general have benefited from the tremendous foresight and energy of Captain Joseph Callaway Lea.

Elvis E. Fleming
Roswell, New Mexico

TABLE OF CONTENTS

LIST OF ILLUSTRATIONS

LIST OF ILLUSTRATIONS cont.

1

INTRODUCTION

Joseph Callaway Lea

Joseph Callaway Lea (1841-1904) was an outstanding man who played a significant role in shaping the West, whose life in many ways reflects the pattern of American history in the nineteenth century. He was particularly influential in leading Southeast New Mexico from a small dot in the cattlemen's and farmers' frontiers to one of the most developed areas in New Mexico Territory. The Pecos Valley/Roswell area in his lifetime became, in many ways, the leading region in New Mexico.

Lea experienced life in a multiplicity of ways. He began life as a farm boy in Tennessee and Missouri, the son of a well-to-do planter and doctor. In his early manhood, he participated in the Confederate guerrilla movement that grew out of the Missouri-Kansas border wars—a prelude to the American Civil War in that area in the 1860s. He served under Quantrill and Shelby, but for much of the war Lea was a captain in charge of his own men in Louisiana.

Moving around the South in the post-war years, Lea, in 1867, married a widow who soon died. More restlessness followed, but, in 1875, he met and married Sallie Wildy, the beautiful daughter of a Mississippi planter. Lea left for New Mexico Territory in search of his fortune. After a few years of ranching in the high country of northern New Mexico, Lea moved his family and livestock to the Pecos Valley. The Leas had two children before Sallie died in 1884. Although Roswell was a small community when they arrived, Lea was the one who actually developed it—thus earning for himself the title of 'The Father of Roswell.'

Lea was much more than the Father of Roswell. He was an important force for peace during the Lincoln County War and played a major role in ridding the country of Billy the Kid. After the violence ended, Lea embarked on a large-scale ranching venture: the Lea Cattle Company. Financed primarily by New York businessman Horace Thurber, the Lea Cattle Company, by some estimates, had the largest ranch with the most cattle of any ranch in Lincoln County and probably in the Territory by 1890.

While the cattle company was formed primarily to produce cattle, it also owned land available for development. To attract farmers to the Pecos Valley, the company embarked on a major effort to develop both the town of Roswell and the farmland of the Valley. They built numerous canals and ditches to irrigate thousands of acres of farmland. They also laid out the original streets and several additions to Roswell and began to sell or even give lots to interested folks who wanted to establish businesses and homes. Economic forces beyond Lea's control gradually, but eventually, brought a close to the cattle company.

Meanwhile, J.C. Lea was the principal mover behind the creation of Chaves and Eddy counties in 1889, but declined to have a county named for him. The incorporation of Roswell, the building of the railroad into Roswell, and the establishment of businesses all felt Lea's influence.

Lea also played the major role in establishing Goss Military Institute in Roswell in 1891. It later became New Mexico Military Institute and continues as one of the premier military schools in the nation.

When Roswell changed its government from a town to a city in 1903, Lea was drafted to run for mayor and was elected to serve until the regular election cycle came around. Unfortunately, he lived only two months after his election.

J. C. Lea received much honor and recognition in his lifetime as well as after his death. He also suffered his share of sorrow, losing two wives to premature deaths, seeing his only son confined to a mental hospital, and watching some of his investments fail. He truly was a remarkable man among men, greatly esteemed by all who knew him. During his adult life, Lea was a guerrilla, a ranchman, a developer, and a patriarch to one of New Mexico's most important communities.

Ancestors of Joseph C. Lea

Joseph C. Lea was a fourth-generation American. He was descended from James Lea, an Englishman born to John Lea at Lea Hall in Cheshire, England, in 1718.

In 1740, James Lea and his brother William, together with another relative of unknown connection, John C. Lea, emigrated to the Colony of Virginia. About 1749, they relocated to North Carolina to a place called 'Leasburg' in their honor. They built an Anglican church there, but later converted to the Baptist faith.

Tradition holds an unusual development for the three Lea kinsmen: projecting that they would become the patriarchs of three separate clans of Leas in America, they decided to spell their names three different ways: Lea, Lee, and Leigh. The Tennessee Leas were descendants of the oldest of the English immigrants, James Lea, who was J.C. Lea's great-great-grandfather.

James Lea married Anne Tolbert. They had four sons and three daughters and apparently lived out their lives in North Carolina.

James' second son, Major Lea, born in 1741, the great-grandfather of J.C. Lea, was probably the first Lea who moved to Tennessee. Major married Elizabeth Herndon, and three children were born to them: Herndon, Abner, and Major, Jr. Major, Sr., served as a magistrate in Grainger County, Tennessee, in 1796. He lived to be 102 and died in Powell's Valley near Cumberland Gap on November 19, 1843.

J.C. Lea's grandfather was Major Lea, Jr., who was born on February 26, 1775. He married Rhoda Jarnagin (born January 26, 1778) in Jefferson County, Tennessee, about 1800. One of their nine children was Pleasant John Graves Lea, born November 6, 1807, probably at Cleveland, Tennessee. Major Lea, Jr., died on August 8, 1821; and Rhoda died on December 21, 1852, near Cleveland.

Pleasant J.G. Lea, J.C. Lea's father, married Lucinda F. Callaway in 1838. Lucinda was the youngest of the four children of Joseph and Nancy Callaway, and she was born in 1821. Pleasant and Lucinda had nine children, the first six of which were born in Cleveland, county seat of Bradley County. Thomas Callaway 'Tom' Lea was born February 1, 1839; Joseph Callaway 'Joe'

Above: Dr. Pleasant John Graves Lea at age 52, in 1858 in Jackson County, Missouri. *Courtesy HCSNM, Photo No. 3338-B.*
Below: Dr. Pleasant Lea's home in what was to become Lee's Summit, Missouri. This drawing appeared in the *Kansas City Star*, January 17, 1932.

'J.C.' Lea, the subject of this discourse, was born November 8, 1841; Franklin Houston 'Frank' Lea was born July 18, 1843; Alfred E. Lea was born about 1845; Elvira Lea was born April 10, 1848; and Mary Lea was born about 1849. The Leas were members of the Methodist Episcopal Church, South.

In the 1960s, a manuscript about the Lea family, written by Claudine Sowell Chandler, Tom Lea's stepdaughter, was placed in the archives of the Jackson County, Missouri, Historical Society. It is the source of some of the facts about Pleasant Lea. He attended Jefferson Medical College in Philadelphia in the 1840's, returning in the spring of 1847 to practice medicine in the Cleveland community. Two years later, he moved to the Missouri frontier.

According to Chandler, the Leas joined a caravan of seven prairie schooners, two wagons, and a carriage loaded with the worldly goods of three other families—one of which presumably was the family of Dr. Lea's brother, Alfred. They set out for Missouri on October 9, 1849. Chandler claims that Mary Lea, who she says was eleven years old, disappeared from camp one night and was never heard from again. This seems to be refuted by the 1850 census, which shows one-year-old Mary with the family in Missouri, although this could be a different child.

After forty-four days on the trail, the Lea family arrived on November 22, 1849, on the site of present Lee's Summit, Jackson County, Missouri. Dr. Lea built a colonial-style house on the road leading to Independence. He became the local doctor for their new community and a very prosperous farmer who owned a few slaves (six in the 1860 census). His brother, Alfred, lived close by and also had several children. The place where Pleasant and Alfred Lea settled their families was known at the time as 'Big Cedar.' In an interview conducted in 1959 and published in 1994 in an article by Joanne C. Eakin in the *Kansas City Genealogist,* Bette Cooper stated, "Dr. Pleasant Lea entered 800 acres, adjoining William B. Howard's land and had the finest home on what is now the McKeighan place ... ." Other sources indicate that Pleasant and Alfred Lea were among the first to settle on the high land that was a safe distance from the deleterious effects of the Missouri River.

Three more children were born to Pleasant and Lucinda Lea at Big Cedar: Caroline 'Carrie' Rebecca Lea in 1852; John Graves Lea in 1854; and Annie Lea in 1856. Annie is not listed in the 1860 census, so she may have died as a young child.

Lucinda F. Callaway Lea died in 1857 and was buried in the family plot at the plantation.

Dr. Lea married again in 1860 when he was in his early fifties. Little is known about this wife. Her name was 'Fanny,' according to the census of 1860. She was about thirty-eight when they married and was a Yankee from Massachusetts.

One important source of information on J.C. Lea is an article by James T. Padgitt in the 1959 yearbook of the West Texas Historical Association. Padgitt's mother was Willie Day Padgitt, J.C. Lea's stepdaughter. Padgitt had the benefit of stories passed down in the family, as well as interviews with surviving Lea relatives; and his article contains information not available in any other source.

According to Padgitt, all of Dr. Pleasant Lea's children attended a common school in their neighborhood. After J.C. 'Joe' had completed the school, he and three of his brothers—Tom, Frank, and Alfred—went to Colorado in 1860. Joe was in charge of a big freight-wagon train, which was no small task for a nineteen-year-old. According to a biographical sketch of Joseph C. Lea in the 1902 *Bronco*, the first yearbook of the New Mexico Military Institute, he had many encounters with Indians on the way and while he was working in Colorado as a prospector, lumberman, and freighter. He visited the little bergs of Boulder and Denver, and then settled on a ranch. He was there until the spring of 1862. Padgitt makes no further mention of the activities of the other brothers.

Lee's Summit

Local historians disagree about the role played by Dr. Lea in the founding and naming of Lee's Summit. Family tradition holds that he may have founded the town and that it was named after him. Local historians there generally do not credit him with founding the town, but some do give him credit for being the source of the name.

Donald R. Hale of Lee's Summit writes that in 1844, the founder of future Lee's Summit, William B. Howard, settled in Jackson County. An article in the *Lee's Summit Journal* on October 13, 1949, states that Howard bought 833 acres of prairie land from the government. According to Hale, a post office was established at Big Cedar in July 1847. When Dr. Lea came in from Tennessee, he built his house near the Big Cedar post office and on the road leading to Independence (today identified as Chipman and Douglas Streets at Lea-McKeighan Park) on the site of present Lee's Summit. Howard built a large, colonial-style house and cabins for his slaves near Big Cedar in 1854.

How the town came to be named Lee's Summit is not clear, but the local consensus is that Howard intended to build a town there and name it in honor of Robert E. Lee, although Lee's fame had not spread far that early in his career. The 'Summit' part of the name is because the area is the highest elevation point on the Missouri Pacific Railroad between St. Louis and Kansas City. Howard was forced by 'Order No. 11' to flee back to Kentucky in 1862 for the duration of the Civil War. Apparently, the area was already being called 'Lee's Summit'; and in Howard's absence, people probably assumed that it was named after Dr. Lea who lived on the site. During the Civil War, the Union troops called the area 'Doctor Lea Prairie,' according to Hale.

J.C. 'Joe' Lea about 1860 at the age of nineteen
when he and his brothers went to Colorado.
Courtesy: HCSNM, Photo No. 3326-A

2

J.C. LEA IN THE CIVIL WAR

Sectional Strife

The passage of the Kansas-Nebraska Act by Congress in 1854 embroiled western Missouri in the slavery issue. That law allowed residents of the two territories, Kansas and Nebraska, to decide by vote whether their states would become free or slave upon statehood. People flocked there from other areas to vote for their views. Clashes between anti-slavery Kansans and pro-slavery Missourians escalated and the area became known as 'Bleeding Kansas.' Jackson County, where the Leas lived, was in the middle of the hostility.

In 1856, a Kansas territorial marshal took a huge federal posse of mostly pro-slavery Missourians to Lawrence, Kansas, where they arrested free-state leaders who had broken the law. Some accounts aver that they sacked the town and killed five free-staters. In retribution, the fanatical anti-slavery advocate John Brown and his followers killed a similar number of pro-slavery settlers—the so-called 'Pottawatomie Massacre'—and left their mutilated bodies as a warning to other supporters of slavery to stay out of the territory. Brown's raid resulted in more civil strife in Kansas, mostly irregular guerrilla warfare carried on by armed bands, many of whom were interested not so much in ideology as in loot and land.

Many people in Jackson County and other nearby Missouri counties, afraid for their lives and property, left the state. Kansas anti-slavery advocates—Jayhawkers and Redlegs (scouts who wore red sheep-skin leggings)—raided along the state line, plundering anything of value, murdering settlers, stealing slaves, and burning buildings.

An important resource for information on the Civil War in Missouri is Donald R. Hale of Lee's Summit, who has written and co-written several books and articles on the subjects. According to Hale, some slavery supporters attempted to organize resistance. Eventually the governor called out the militia to quell the hostility in 1858. However, the Kansas-Missouri border clashes started up again when the Civil War began in 1861.

A real civil war broke out in Missouri between the Unionists of Nathaniel Lyon and the secessionists led by former governor Sterling Price. Missouri remained a Union slave state, which was not so much the will of the people of Missouri as it was the dominating forcefulness of President Abraham Lincoln's government. Even so, thousands of Missourians fought for the Confederacy.

As early as the winter of 1860-61, support for secession was rather strong in Independence—the county seat—and the rest of Jackson County except for Kansas City. Upon the start of the sectional war, Jackson Countians formed Confederate army companies. The Missouri State Guard favored secession, but Federal troops ran them out and occupied the state. The Northern soldiers apparently believed that almost everyone in Missouri was loyal to the Confederacy and treated them accordingly, Hale states. This animosity toward Southern sympathizers, both real and imagined, resulted in abuse of the civilian population by the Union troops.

Guerrillas operating independently of the US Army perpetrated many of the excesses against the population. Among the anti-slavery Easterners who had settled in Kansas and who became fanatical abolitionists were Dr. Charles 'Doc' Jennison, Kansas U.S. Senator Jim Lane, and James Montgomery. Acting on behalf of the Union, these men and their followers occupied much of western Missouri in the summer and fall of 1861. The guerrilla war that ensued was largely their doing, Hale states.

These three abolitionists had already raided Missouri and harassed pro-slavery settlers in Kansas Territory before the war. The war provided new opportunities, not only for fighting slavery, but also for plundering the goods of the settlers, both Confederate sympathizers and Union sympathizers. This ruthless harassment caused many in Missouri to become disenchanted with the Union. Young men in Jackson County and nearby areas were ripe for recruiting into a Confederate guerrilla force to counter the Jayhawkers. All they needed was a leader.

William Clarke Quantrill

William Clarke Quantrill answered that leadership need. Quantrill was born in Ohio in 1837; he spent a few years in the West as a schoolteacher and gambler. Some accounts refer to him as 'Charles William Quantrell' or 'Charley Hart,' which names he may have used at one time or another. According to conventional accounts, he landed in Missouri in late 1860, while fleeing from charges of horse stealing and murder in Utah.

Quantrill jumped into the sectional strife in Missouri and Kansas as a fanatical, anti-abolitionist, Confederate sympathizer. He claimed to have watched the Jayhawkers kill his brother, but scholars now believe he made up that story in order to elicit Confederate sympathy for his cause. By the end of 1861, he led one of the most feared fighting outfits of the Civil War. His style of guerrilla warfare terrorized some civilians, but was generally directed at Union military personnel. The army of irregulars that Quantrill raised included 'Bloody Bill' Anderson (whose company included Jesse James, according to Hale), Cole Younger and his brothers Jim and Bob, and Frank James. Quantrill's contingent became an important force that harassed Federal troops and Union sympathizers in the Missouri/Kansas border area, often fighting against Kansas Jayhawkers who raided into Missouri from Kansas.

Cole Younger's father, Henry, was a substantial, prominent man in the Jackson County area. He had served as a state legislator and a Cass County Judge, served as mayor of Harrisonville, and owned a number of businesses and much land in the area. Most important, he was a slave-owning opponent of secession. Those sympathies changed, however, in 1861, when Doc Jennison made his first raid through Jackson and Cass counties. Jennison took all the horses and rigs from Younger's livery stable; and Kansas raiders continued to appropriate his horses and equipment during the following year. As a result of the animosity toward Federal troops on the part of the Youngers, Hale writes, Cole Younger joined Quantrill in

October 1861. In Younger's autobiography, he states that he and his brother-in law, John Jarrette, joined 'Capt. Quantrell's company.' However, it apparently was several months before he was formally enrolled.

The intent here is to examine those aspects of the Civil War that involved J.C. Lea. This is not, therefore, a comprehensive history of the Civil War in the West, the guerrilla movement, W.C. Quantrill, or the state of Missouri. While Lea was connected with numerous skirmishes and actions, it should be remembered that there were countless other actions going on at the same time all over the West—many of them of more consequence than those in which Lea was a participant.

Unfortunately, neither the National Archives nor the Missouri State Archives has the military records of Joseph C. or Frank Lea. Since Missouri was not a Confederate state, the records of only about half the 30,000-40,000 Missourians who served the Confederacy have survived. J.C. and Frank Lea's records are not among them. Fragments of J.C.'s military service can be extrapolated from memoirs and books written about his military units, but it is not possible to reconstruct an exhaustive study of his individual career.

There is little doubt that Joseph C. Lea and his brother, Frank, rode with Quantrill. Indeed, Carl W. Breihan, in his decidedly anti-Quantrill book *Quantrill and his Civil War Guerrillas* (1959), gives a "verified and complete ... authentic list of the 296 men who rode with ... Quantrill at one time or another." Joseph C. Lea is on the list: "Joe Lea (died in Roswell, New Mexico in 1904)," but Frank's name does not appear.

It is not clear when the Leas joined the guerrillas. The initial members of Quantrill's band were local farm boys from Jackson County and adjacent counties. Padgitt avers that the Younger brothers and the Lea brothers were cousins. In any case, Cole Younger was to attract attention and to be closely associated with the Leas.

According to Padgitt, J. C. Lea was in Colorado when the Civil War broke out and caused a renewal of the Kansas-Missouri Border War. Problems were severe back in Jackson County for J.C.'s father, Dr. Pleasant Lea. His new wife Fanny was a Yankee, which made him suspect to some Southern sympathizers, Padgitt claims. However, he was a slave owner and a Southerner by birth, so that caused the Union sympathizers to view him with a suspicious eye. Mrs. Lea took her two young stepdaughters and fled to the safety of her parents' home in Massachusetts, Padgitt writes.

Death of Dr. Lea

Some of the information written by local historians and placed on file in the public library in Lee's Summit (without bibliographical information) indicates that Dr. Lea was killed because his sons and nephews were in the Confederate service; or, more specifically, he had one or two sons "... who had been known to be riding with Quantrill ... ." In a letter to Donald R. Hale of Lee's Summit on February 23, 1973, Don Hoog of Cincinnati avers that J.C. and Frank's younger brother, Alfred E. Lea, also was in the guerrilla movement.

The account given in *History of Jackson County, Missouri* by W.Z. Hickman (1920) indicates that Dr. Lea suffered harassment similar to that inflicted upon Henry Younger. Federal officials, in trying to control the uprisings and violence in Missouri, issued a number of orders. Anyone suspected of being a Southern sympathizer was sub-ject to arrest, incarceration, fines, banishment, or death. Their homes and farms typically were looted by Jayhawkers or unknown outlaws.

There are several versions of how Dr. Lea met his death at the hands of Jayhawkers or Union soldiers. One unidentified source in the Lee's Summit public library states:

> Dr. Lee [sic] lived and owned many acres of land in and around the station on the Missouri Pacific Railroad ... . His land was considered to be of great value. And, for no apparent other reason, than the fact that Dr. Lee had sons and some nephews in the Confederate service, he was persecuted to the extreme. When the men came that final fateful day, it appears, Dr. Lee, perhaps believing in a just and merciful God, who would stop the bullets, allowed himself to be led out into his yard, with no offer of resistance, and there he was shot full of lead. His wife and family looking on.

A tribute to Dr. Lea that was published during the Lee's Summit centennial in the 1960's, declares:

> He was assassinated by Civil War soldiers near the railroad tracks. It was never known whether it was the North or South who was responsible for the death. All that is known is that a messenger arrived at the Lea home with the information that his son, who had been known to be riding with Quantrill, was injured and awaiting medical aid at the Bayles home [originally the Hargis house] on the other side of town.

Hale, an expert on Quantrill, has studied the history of that area for many years. In his manuscript, "Lee's Summit in the Civil War," Hale writes:

> ... One day in 1862, Doctor Lea went to the Josiah Hargis house. The Hargis home was the only home on the [present] Lee's Summit downtown site at that time. It stood at about [present] Fourth and Jefferson and was about a mile from Doctor Lea's home.... Lea had two sons who were with the guerrilla leader William Clarke Quantrill.
>
> On his way some Iowa soldiers who were stationed at Independence stopped him. The Doctor was taken to the vicinity of the present train depot (the Chamber of Commerce) and both his arms were broken, probably trying to gain information about the guerrillas. They then shot him and his home was burned, along with thirteen other homes that night. After Doctor Lea was killed, Hargis and several others took his body in a big wagon to his farm and buried it by the side of his first wife who had died previously....

Another source indicates that it was Doc Jennison's Jayhawkers who killed Dr. Lea.

These accounts suggest that Dr. Lea was killed because his sons were with Quantrill. Some accounts, however, take the position that his sons joined Quantrill because the Union men had killed their father. For example, in her old age, Ella Lea Dow, wrote to a Mrs. Welch and states that J.C. and Frank witnessed the death of their father in December 1861. "My father, Joe, and Uncle Frank left at once to join anyone they could find ... ."

Another version that appeared in the *Kansas City Times* on March 10, 1966, has it that Dr. Lea's family was away, as were many of their neighbors. The Kansas Redlegs came one night and pretended to seek road directions in order to get Dr. Lea to come out of his house. They not only shot him but also shot the only remaining slave still on the place. They took some furniture and burned the house. "After this," the article claims, "two of the doctor's sons, Frank Lea and Joe Lea, returned home and joined the Quantrill guerrillas. Three other brothers, Alfred, John, and Tom Lea, Sr., joined the Confederate army ... ." How John Lea made this list is not clear, since he was only six years old at the time of the 1860 census.

Another source about the Leas is *Buckboard Days,* a book by Sophie Poe (1936). Mrs. Poe lived with J.C. Lea's family in Roswell for a few years in the early 1880s, so she was in a position to hear the stories of Lea's past. She agrees with the *Kansas City Times* article about the events of Dr. Lea's death, but she adds other events between that and J.C. and Frank's joining Quantrill. According to Poe, the Lea brothers tried to stay neutral; but that was not possible because they were considered Southerners by the local Union sympathizers.

As Joe and Frank harvested corn in their father's field, Poe states, a group of Kansas soldiers arrived and took them prisoner. The following day, they were lined up in front of a firing squad, along with eighteen others. The officer in charge was an old acquaintance. He whispered to them, "When I give the command to fire, you both duck and run like hell. Don't stop either, until you join the Confederates!" They were able to get away in the chaos that ensued. It was not long before they joined Quantrill's guerrillas; but Poe, like most other sources, omits any reference to Quantrill with respect to the Leas.

A biographical sketch of Lea's life appeared in the February 12, 1904, issue of the *Roswell Record* upon his death; his daughter, Ella, probably furnished it. The article states that Joe and Frank Lea were arrested in December 1861 and that they made "... their escape just before all the other captives of that raid were shot down ... ." In reference to the 'Kansas borderers,' the article calls them "… those delectables who were too cowardly to go to the front and who showed the manly valor in murdering non-combatant old men, insulting and harassing helpless women and children and stealing and burning property."

In Defense of the Guerrillas. Much as modern-day descendants of Georgia Confederates harbor deep hostility about U.S. General William T. Sherman's 'March to the Sea' in 1864, so some present-day residents of Missouri have similar feelings about the guerrilla warfare of the Kansas Jayhawkers in their area during the Civil War. Those in Missouri who are defenders of Quantrill believe he was unjustly labeled as a vicious killer by contemporary Civil War reporters and by subsequent historians who maintained an anti-Southern bias.

Indeed, *The No Quarterly*, the newsletter of the William Clarke Quantrill Society, in its January 1999 issue, carries a letter expressing these views from a descendant of one of the Jackson County guerrillas. It states, in part:

> I too am interested in seeing the historical record set straight as to the alleged 'depredations' of W.C. Quantrill perpetrated as a 'guerrilla' fighter during the War for Southern Independence. In short, from what I have read of Quantrill, I find his acts to be no worse, and certainly, less in extent that those committed by the federally-revered Gen. William T. Sherman.
>
> In comparing the two men's warring acts, Quantrill's raids seem greatly exaggerated and over-emphasized. Possibly a reason for Sherman's terrorist acts being acceptable to Yankee historians was that his similar, far larger-scale depredations were done in the name of a federally-approved battle campaign of the Union Army, while Quantrill's smaller-scale depredations were done largely and independently of the Confederate Army, as a 'guerrilla' fighter … .

These sentiments seem to express the views of the present-day defenders of Quantrill. However, these views are not of recent origin. Major John N. Edwards, Shelby's adjutant, published his classic *Shelby and His Men, or, The War in the West,* in 1867, only two years after the Civil War ended. He defends the guerrilla organization of Missouri:

> Composed of men driven to desperation by the unceasing persecutions of Federals and militia, they had been outlaws and hunted from county to county as rabid animals. Their scalps commanded a premium, and their mutilated bodies were denied sepulture ... .

Edwards explains how the atrocities committed against relatives of Quantrill, Younger, and others motivated them to join the guerrilla movement. Further, he contrasts the public outcry over the actions of the Confederate guerrillas with the lack of public outrage when Union elements committed similar acts, noting that "... white plethoric eyes rolled horrified heavenward..." at Quantrill's offenses, but "... not a pious Puritan marred the platitude of his sleek, meek face by frowning at the enormities in Jackson, Bates, Cass, and Johnson counties in Missouri."

These passages offer some insight into why some patriotic, law-abiding, honest citizens gave their support to the Confederacy and dared to ride with guerrilla leaders such as William C. Quantrill.

Action With Quantrill

Cole Younger's father met a fate similar to that of Dr. Lea. On July 20, 1862, Henry Younger was ambushed and killed five miles from Independence. Hickman writes:

> ... this and other insults that had been offered to the Younger family brought to the front one of the most daring and dangerous characters that ever drilled beneath the black flag. Coleman Younger, ... while beholding the agonizing tremors of his delicate mother and sisters over the dead body of his dear father, made resolutions the faithful carrying out of which has cast a shadow over his father's family and good name, made hundreds of widows and fatherless children, and scattered forgotten graves over the entire portion of western Missouri and eastern Kansas ... .

In his autobiography, Younger says that in July and August 1862, Colonel Upton Hays recruited a regiment of farm boys for the Confederate Army in Jackson County for the specific purpose of dislodging U.S. Colonel James T. Buell from Independence. Younger helped Hays recruit young men and was himself formally enrolled in the CSA Army, along with about three hundred others who were sworn in at Luther Mason's farm. Younger states he was in Captain John Jarrette's company in Colonel Upton B. Hays' regiment. The outfit was officially designated as the Second Missouri Cavalry, according to Hale, and it became a part of Joseph O. Shelby's Sixth Missouri Cavalry Brigade. It seems likely that this was the date of the official enlistment and the outfit which Joe and Frank Lea joined. On August 11, in the Battle of Independence, the farm boys led by Quantrill captured the town from Buell's 450 Union troops, Hale relates in his booklet, *Civil War and Reconstruction in Independence, Missouri.*

Because of constant changes in assignments, promotions of officers, reorganization of military units, and the vicissitudes and fortunes of war, it is impossible to trace the movements of J.C. Lea during the Civil War. The *Roswell Record* (February 12, 1904) states:

> ... Lea grew to manhood a hard-working, energetic farmer boy of simple tastes, and who viewed the internecine struggle then just beginning as something at a distance that did not especially concern boys of his age .... He entered the service a farmer boy without any training and declined a colonelcy before the third year of his service had expired.

It appears that Lea began with Quantrill, then shifted to Shelby's Brigade, and, to some extent, was back-and-forth between the two outfits for a while. Lea operated under other officers after 1863, but he became an officer with considerable rank and authority of his own.

As a rookie guerrilla under Quantrill, Padgitt allows, J.C. Lea was motivated by a desire for revenge. He "quickly completed the school of the bushwhacker" and learned from Quantrill to always count the cost of a battle, determine the odds, retreat if you cannot win, but always strike back furiously, fight desperately, and leave nothing alive on the battlefield. Lea emerged as a seasoned guerrilla and bushwhacker leader with "extraordinary resources and cunning."

He was a fearless young man who had experienced many close calls in the Border War. Padgitt describes him as powerful, over six feet tall, with a "wild, dashing air that always distinguished him in any company, and had a face altogether more pleasing than disagreeable."

With strong opinions on the part of some Missouri historians in opposition to conventional historical accounts of Quantrill, it does not seem possible to devise a narrative of guerrilla activity during the Civil War which will satisfy all viewpoints. Quantrill's detractors maintain that Quantrill, leading his band of guerrillas, became a plague to both Kansas and Missouri. They sometimes fought skirmishes with Union soldiers; but mostly they were busy raiding, pillaging, and burning farmsteads and small towns believed to be inhabited by Union sympathizers, sometimes killing the civilians there. Proponents of Quantrill hold that he was only responding to similar depredations by Kansas Jayhawkers and his actions were legitimate military operations against Union troops, not civilians.

In any case, the U.S. Commanding General for Missouri, Henry B. Halleck, on March 13, 1862, formally declared that all guerrillas were outlaws. This meant, according to Donald R. Hale in his booklet, *Jackson County and the Civil War,* that any guerrilla was subject to hanging if captured.

After the Battle of Independence on August 11, 1862, an important battle was fought at Lone Jack on the 16th. Union troops thought they were fighting Quantrill and so fought furiously and refused to surrender until they learned that Quantrill was not there—although many of his men were. According to Hale, Quantrill was in Independence that day securing military supplies.

Colonel Hays withdrew to the Little Blue River and continued his recruiting. A fight with Federal forces there was a draw. Whether or not the Leas participated in these activities of Hays and/or Quantrill is unknown, but it can be assumed they were involved in some of the actions.

Quantrill assembled his forces at Sibley on October 5, and on the following day they ambushed a Federal force that was hunting the guerrillas. Quantrill's men scattered and took off for Independence, and the Union troops gave up.

Hale states that Quantrill moved without resistance throughout Jackson County between October 8th and 16th with about 150 guerrillas. They soon relocated in Arkansas to the south. The next spring, they were back in Jackson County. Quantrill's men—including Cole Younger and probably the Lea brothers—took the *Sam Gaty,* a steamboat at Sibley Landing, on March 26, 1863. They killed three passengers, captured twenty slaves, and dumped twelve wagons and a hundred sacks of flour into the river.

J.O. 'Jo.' Shelby

Joseph O. 'Jo.' Shelby, as a captain in the Missouri State Guard, was frequently associated with General Sterling Price, the first head of that outfit. Price had been the military governor of New Mexico Territory in 1846-47 during the military occupation and then served as governor of Missouri in the early 1850's. When the Civil War started, Federal armies were sent into eastern Kansas and central Missouri. On August 10, 1861, there was a major conflict at Wilson's Creek in which Shelby participated. After that, Missouri was a state "divided against itself."

At the end of 1861, the Missouri State Guard reorganized into the Confederate service, and Price became a Confederate officer. Price and Shelby fought together at the Battle of Pea Ridge (Elkhorn Tavern) in Arkansas (March 1862). Shelby also was at St. Charles and Devall's Bluff in that state. In the spring of 1862, Shelby transferred to the regular Confederate Army. In a battle at Newtonia on April 30, 1862, Shelby's cavalry and three thousand Indians under Douglas H. Cooper fought the 1st Brigade of the Army of Kansas and the 4th Brigade of the Missouri Militia Cavalry. The Union troops had fifty killed, eighty wounded, and 115 missing. Shelby and Cooper had 220 killed and 280 missing.

Missouri Raid, 1862. Shelby joined General James E. Rains in June 1862, at Van Buren, Arkansas, in preparation for invading Missouri. In an unusual show of support and loyalty, his men swore they would never lay down their arms until the war was over.

In the summer invasion of Missouri, Shelby distinguished himself in the Battle of Carthage. He recruited a cavalry regiment of a

General Joseph O. 'Jo' Shelby, commander of the Confederate Sixth Missouri Cavalry Brigade, in which J.C. and Frank Lea served. From an engraving by A.H. Ritchie. *Courtesy: State Historical Society of Missouri, Columbia.*

William Clarke Quantrill, Captain of Cavalry Scouts in Shelby's Brigade under the Partisan Ranger Act. He was the Confederate guerrilla leader under whom J.C. and Frank Lea started their military service. *Courtesy: State Historical Society of Missouri, Columbia*

a thousand men. As a reward he was promoted to colonel, organized the Sixth Missouri Cavalry Regiment—which he liked to call the 'Iron Brigade,' and took command at Newtonia under Major General Thomas C. Hindman in the fall of 1862. Shelby's cavalry force was one of two brigades comprising CSA General John S. Marmaduke's division, and it continued to be identified with the campaigns of Price.

After the Battle of Independence on August 11, 1862, Quantrill became Captain of Cavalry Scouts in Shelby's Brigade under the Partisan Ranger Act; and his men were mustered into the regular Confederate service. Accounts indicate that Quantrill and some of his men maintained a considerable degree of independence, moving in and out of Shelby's unit as the situation required—a pattern which seems to apply to J.C. Lea. Quantrill disbanded his unit for short periods from time-to-time in order to foil pursuers.

The armies that met in the fall of 1862 at Newtonia, in the southwestern corner of the state, were quite large. The Union officers discovered that Shelby, Cooper, and Hays were dug in behind the rock walls of the town in an invulnerable position. U.S. General John M. Schofield came to personally conduct the attack on the Confederates. He was highly successful, and sent the Southerners beating a hasty retreat on October 4. Hays died in the fighting, but it is not known if his death affected the status of Younger and Lea. Cooper fled into Indian Territory; and Shelby led his Iron Brigade into the Boston Mountains in southwestern Arkansas. Every organized Confederate unit had been driven out of Missouri.

Shelby enhanced his military reputation during the retreat by devising a rear-guard tactic to protect the retreating armies from enemy fire. The tactic was so successful that it was used several times later, according to Alvin M. Josephy, Jr. in *The Civil War in the American West* (1991).

Battle of Prairie Grove. Shelby spent the fall and winter near Huntsville, Arkansas, making forays against U.S. General James G. Blunt. Marmaduke, the top commander in the battlefield, had a cavalry which included Quantrill's bushwhackers as well as Shelby's cavalry. At one point, Shelby himself was captured and then rescued. In a notable action at Prairie Grove, on December 7, 1862, Shelby waylaid three companies of Blunt's Sixth Kansas Cavalry. This was part of a much larger, hard-fought battle in which each side had about ten thousand soldiers. Many of the Arkansas Confederate conscripts were loathe to fight. Blunt bombarded them for an hour and then rolled up Marmaduke's flank. Shelby's cavalry drove them back. Marmaduke called for a truce to bury the dead, then during the night high-tailed it for Arkansas. The Confederates had suffered 1,317 casualties; and the Federals had 1,251. Marmaduke's division virtually fell apart during the retreat. Only about five thousand ill, barefooted, and hungry men remained.

Fortunately for the Confederate cause, Marmaduke's cavalry division—including Shelby's Brigade—managed to hold together. On December 31, 1862, Marmaduke's Confederates plunged into Missouri to capture the outer defenses of Springfield. Shelby was sent to cut the railroad supply line coming from Rolla. Furious resistance in the Springfield campaign from the Missouri militia, citizens, and convalescents drove the Rebels back. There were fourteen Union deaths and 144 wounded, while Marmaduke lost forty killed and 206 wounded and missing. Marmaduke's army retreated in the snow and ice to Batesville, Arkansas, and rested there until spring.

Marmaduke's Missouri Raid, 1863. In the spring of 1863, Marmaduke and Shelby attempted another raid into Missouri, this time to Cape Girardeau in Southeast Missouri. Shelby's forces attacked the town on April 26, and engaged in hard fighting. Ultimately the Confederates retreated and fought several battles with the Union troops as they moved south out of Missouri. Shelby's brigade then encamped near Augusta, Arkansas.

Battle of Helena. Thinking they saw a chance to interrupt U.S. General U.S. Grant's communication lines, the Confederates attacked Helena, Arkansas, on July 4, 1863. They did not know Vicksburg had already fallen. Shelby was wounded in the failed action when a rifle ball shattered his wrist, but he saved his battery.

Quantrill's Raid on Lawrence, Kansas.

Hale writes that U.S. General Thomas Ewing issued his

‘Order No. 10’ on August 18, 1863, ordering that women heads-of-families who were helping the Confederate guerrillas be removed from the state of Missouri. Many facilities where the women were incarcerated were filled beyond their capacity. One such building in Kansas City collapsed from the weight after Union troops undermined it, killing four and injuring others.

This collapse was the spark that set off Quantrill, who had relatives in the collapsed building. Lawrence, Kansas, located not far west of Jackson County, was the Jayhawkers’ headquarters. Quantrill decided to attack their home base.

Quantrill’s raid on Lawrence is considered by Union historians to be one of the most disastrous and dastardly events in the West during the Civil War. Confederate historians believe the raid was a justifiable military action. The force with Quantrill was estimated to be as many as 450 men. They crossed the Kansas River on August 20, near Aubrey. A hundred Union troops watched them cross; however, they did not take up the chase in time to stop Quantrill from reaching Lawrence.

Crossing fifty miles of the countryside in Kansas without prompting an alarm, Quantrill reached Lawrence early on the morning of August 21, and proceeded to attack the large contingent of Jayhawkers camped in the middle of town. One of his main targets was U.S. Senator James H. Lane, a free-state champion in Kansas and a leader of an anti-slavery Jayhawker army. Dashing through a cornfield in his nightshirt, Lane escaped. Others were not so fortunate. Estimates of people killed by Quantrill’s men run from 150 to 183. Confederate historians claim the casualties were all men, while Union historians allege that women and children were also killed. More than two hundred residences, hotels, and stores were destroyed.

Historical evidence indicates that J C. Lea was a participant in the Lawrence raid and, in fact, was wounded there. In their monumental reference book, *Branded as Rebels* (1993), Donald R. Hale and Joanne Eakin provide a thumbnail sketch of Joseph C. Lea: “With Quantrill .... Wounded at Lawrence on August 21, 1863 ... .”

On August 25, 1863, USA General Ewing issued ‘Order No. 11’ which directed all residents of Bates, Cass, and Jackson counties, and some of Vernon County, to leave their homes unless they could prove to the closest military commander that they were loyal to the United States. If they could, they were allowed stay at the military posts or move to Kansas. According to the *Kansas City Star* of January 17, 1932, Order No. 11 was reportedly written in Dr. Pleasant Lea's house, which apparently had been restored following the burning in 1862.

Shelby's 'Great Raid Into Missouri'

Colonel Shelby recovered enough from the wounds he received in the Battle of Helena to take command of his men. On September 22, 1863, he took his Sixth Missouri Cavalry Brigade and a battery of light artillery and embarked upon the greatest accomplishment of his career: an invasion of Central Missouri. Daniel O'Flaherty in *General Jo Shelby: Undefeated Rebel* claims it was the longest and most daring raid of the Civil War. Shelby started with about six hundred men, and bushwhacking bands joined them until the force totaled between 1,500 and 2,000.

Another Captain Lea. Edwards, in his classic *Shelby and His Men,* makes several references to 'Captain Lea' in his descriptions of Shelby's military operations; but he never gives a first name. Previous New Mexico historians have assumed that all of Edwards' references to 'Lea' mean J.C. Lea. This is the case in James D. Shinkle's *Fifty Years of Roswell History, 1867-1917* (1964) and William E. Gibbs and Eugene T. Jackman in *New Mexico Military Institute: A Centennial History* (1991). However, it is likely that the Captain Lea mentioned by Edwards was not always J.C. Lea; but it is not possible to be certain. Probably any references to a Captain Lea by Edwards in Shelby's operations after the fall of 1863 are not referring to J.C. Lea.

Some of the references to 'Captain Lea' by Edwards may well be about Captain Thomas H. Lea, who enlisted in Newton County, Missouri, on September 12, 1862, at the age of thirty years. He was immediately elected captain of Company 'A,' of the Third Regiment (Thompson's) Missouri Cavalry, which was a unit of Shelby's Sixth Missouri Cavalry Brigade of CSA General Sterling Price's

Division of the Confederate Army. Some of Thomas H. Lea's records are under the name of 'J.H. Lea,' but there seems to be no doubt that this is the same man. Whether or not he was related to J.C. and Frank Lea is unknown; but it is certain that he was not their older brother, Thomas Callaway Lea. Their uncle Alfred Lea of Jackson County had no son by that name, and it is not known if this Thomas H. Lea was related through some other branch of the Tennessee Lea family. One account states that Thomas C. Lea was also a captain in Shelby's Brigade, serving in Smith's Regiment.

Baxter Springs, Kansas. J.C. Lea was a part of Shelby's 1863 raid as a member of Quantrill's band. Two months after the Lawrence raid, Quantrill defeated a small group of Union cavalry at Baxter Springs, Kansas. This operation apparently was a 'sideshow' of Shelby's raid. Union historians claim that Quantrill captured seventeen non-combatants, mostly members of a brass band, and summarily executed them. Quantrill's defenders, on the other hand, hold that the band members were soldiers and were killed in the fighting.

Boonville: 750-Mile March. On October 1, 1863, Shelby took Neosho in the southwestern corner of Missouri, then moved north. Other towns in his path included Bowers' Mill, Tipton, and Humansville. An important milestone in Shelby's invasion was his attack on the town of Boonville in north-central Missouri, which he reached after traveling 750 miles from his starting point. Edwards notes that the Federal troops at Boonville were laughing and singing, wholly unaware of what was about to hit them. "Captain Lea struck first, and the foremost squadron went down almost to a man. Another pushed on and shared the same disaster from the cool riflemen under Toney ... ." According to Jay Monaghan in *Civil War on the Western Border, 1854-1865* (1955), the mayor of Boonville invited the raiders to enter the town peacefully. At this time, Shelby began recruiting soldiers from north of the Missouri River.

Lea at Marshall. At nearby Marshall, Shelby encountered State Militia General E. B. Brown. "Upon the side of the stream nearest Brown," Edwards writes, "were stationed the cool and intrepid Lea and his company, with orders to fire one volley and retreat in disorder, as if demoralized and panic-stricken." The ploy worked, and the result was the slaughter of 111 Federal troops in the ambush, while only one Confederate was wounded.

Lea at Carthage. As Shelby marched south back toward Arkansas, he went west of Clinton and east of Warrensburg. He stopped to feed the horses and to allow a detachment to burn a bridge over the LaMine River. South of Clinton, some Kansas cavalry attacked Shelby's rear, but the resistance they met was so fierce they left. In a day and a half, Shelby marched a hundred miles to Carthage, west of Springfield. During that time he had stopped five times to feed the horses and had been attacked in the rear twice. Carthage was relatively safe ground, and they camped near there for a good night's rest. Some of the command that he allowed to camp in the town were surprised by U.S. General Thomas Ewing's advance guard and driven out.

Shelby heard the commotion and ambushed Ewing's army. He sent five companies, including one under a Captain Lea, to hold Ewing in check. They stopped him for an hour, forcing him to go into line of battle and to activate his artillery. Shelby's well-mounted brigade broke quickly into column and galloped southward in triumph. Ewing abandoned the pursuit, so Shelby was able to move unhurriedly to the White River and encamp near Berryville to rest his command.

A Union account of Shelby's Raid states that Shelby was defeated on October 13, 1863, after a sharp fight of several hours with Brown and Ewing. They chased Shelby for over two hundred miles until his force fell apart in the mountains of Arkansas.

Lea's 'Ragged Wound.' At one point during the retreat, Edwards describes a "forest respite" which gave the men a breather. He describes what each of the officers was doing while they rested, such as smoking, arguing, or reading. One Captain Lea was wounded in the campaign; and Edwards describes Captain Dickey "... binding up the ragged wound of Captain Lea, and pouring confidence into his comrade's heart."

Results of Shelby's Raid. One Union account states that Shelby was routed from Missouri, so the western part of the state was spared the raiding of both Shelby and Quantrill for the rest of 1863. Confederate sources tell a different story. During their retreat, Shelby's men exhibited a sense of flair by wearing captured Union blue uniforms and a plume of red sumac in their hats. They went to Camden, Arkansas, for the winter. Shelby's raid enhanced his reputation throughout the Confederate military command. Confederate officers believed that the raid kept ten thousand Union troops from reinforcing U.S. General William S. Rosecrans following his defeat at Chickamauga, Georgia.

More obvious were the physical effects of the raid. Altogether, Shelby's forces fought forty-seven skirmishes and battles and captured twenty garrisoned towns. They destroyed eleven forts and blockhouses, according to John C. Moore in *Confederate Military History, Volume IX: Missouri.* The men had marched 1,500 miles in forty-one days, destroying $1,000,000 worth of Union supplies and $800,000 worth of railroad track, water tanks, bridges, and telegraph wires. While losing 150 men, they killed or wounded six hundred Union men and captured and paroled five hundred more. They seized six hundred guns, six hundred revolvers, three hundred wagons, and $50,000 worth of ordnance—which they destroyed. The men also brought back eight hundred new recruits from the northern part of the state, as well as six thousand horses and mules.

Shelby was received as a conquering hero when he returned to Arkansas. He infused a new courage and confidence into the Trans-Mississippi Confederate Army by demonstrating what a bold leader with determined men could achieve. In recognition of his tremendous accomplishment, Shelby was promoted to brigadier general, according to the Sons of Confederate Veterans.

Break-Up of Quantrill's Guerrillas

Union forces sought retaliation for Quantrill's Lawrence raid. In a 'scorched earth' policy designed to deny the use of resources to the guerrillas, they forced the residents of Jackson and three other Missouri border counties out onto the open prairie while the Jayhawkers plundered or destroyed everything the Missourians had not taken with them. The Confederate forces, including Quantrill's men, tried to visit retribution upon the Union troops for the Jayhawkers' actions. Some six thousand Federal troops routed the Confederates, however; and Quantrill disbanded most of his unit and left for Louisiana with 150 men. After this point, any mention of a 'Captain Lea' in Edwards' book probably refers to Captain Thomas Lea.

CSA General Edmund Kirby-Smith, who took command of the Trans-Mississippi Department on March 18, 1863, and was in command of Confederate forces in Louisiana, quickly informed Quantrill's men that they could not join his forces nor even stay in the state, according to Padgitt. Quantrill moved to Mineral Creek, fifteen miles northwest of Sherman, Texas, and encamped for the winter. Dissension among the guerrillas eventually resulted in Quantrill's outfit dissolving into several smaller bands. One of the most notorious was that led by Quantrill's lieutenant, 'Bloody Bill' Anderson.

Lea and Younger Still Together. According to Younger, he and J.C. Lea were still together in the fall of 1863, apparently in North Texas with Quantrill. They were both in a company under Captain John Jarrette, Younger's brother-in-law. It is not known whether this is the same company into which Jarrette and Younger had enlisted in August 1862, under Colonel Hays. Jarrette left to rejoin Shelby on the Red River, and nineteen-year-old Cole Younger was elected captain. Twenty-two-year-old Joe Lea was chosen as First Lieutenant and Lon Raily as Second Lieutenant. These offices were swapped around once more in the next few months as Jarrette came back, then left again.

According to Younger's autobiography, when he became captain of the company again, he scouted for CSA General Henry E. McCulloch at Bonham, Texas. Later, Younger, Jarrette, and others reported to CSA General E. Kirby-Smith at Shreveport. He sent them on an expedition to get rid of cotton thieves and speculators in the Mississippi river bottoms. They charged one such convoy and captured $180,000 in cash plus the wagons and cotton, Younger states. He led other charges at Bayou Monticello, Goodrich's

Thomas Coleman 'Cole' Younger, comrade-in-arms of J.C. and Frank Lea in the Civil War. Younger later became a notorious outlaw. This photo was taken at Madelia, Minnesota, September 21, 1876. *Courtesy: State Historical Society of Missouri, Columbia.*

Captain Jason W. James, another Civil War comrade-in-arms of J.C. and Frank Lea. *Courtesy: HCSNM, Photo No. 4117.*

Landing, and Omega to end the problem in that section. Younger apparently at this time was under Shelby's command.

The breakup of Quantrill's forces probably was the time when J.C. and Frank Lea and many others separated themselves from both Quantrill and Shelby. The Leas were no longer involved with Quantrill as he persisted in his operations against the Union. He led thirty-three guerrillas into Kentucky in May 1865, where he was fatally wounded in a skirmish with a small, irregular Union force.

Captain Jason W. James

Glimpses of J.C. Lea's activities during the last two years of the Civil War can be seen in the accounts about his comrade-in-arms and later fellow resident of Roswell: Captain Jason W. James, who was born in Lexington, Missouri, in 1843. A letter in the files of the Western Historical Manuscript Collection at the University of Missouri-Columbia asserts that James "... fought with Quantrill in the Kansas and Missouri Wars, and later served with distinction in the Confederate States Army under Generals Joe Shelby and Kirby Smith ... ." Francis M. Ward of Tallulah, Louisiana, wrote the letter to B.J. George, Sr., of Kansas City on August 10, 1959. George was the son of Hiram George, who rode with Quantrill.

In James' *Memorable Events in the Life of Captain Jason W. James*, he states that when the Civil War started in the spring of 1861, he signed up in the Missouri State Guard, Captain C. J. Kirtley's company. The purpose of the company was to drive out the Jayhawkers and guard the border. He was involved in a number of skirmishes during his six-month term of enlistment. James re-enlisted in St. Clair County, Missouri, on December 10, 1861, this time in the regular Confederate Army. He was in Company H of the Missouri Light Artillery (CSA). The records show that this battery was originally Captain Churchill Clark's battery, then became Captain Houston King's battery upon Clark's death, and subsequently became "Captain James L. Farris' battery (Clark Artillery), Missouri Light Artillery (Confederate)." James was promoted to corporal on April 28, 1862.

In his book, James tells of the many battles he fought under Price and Shelby, including the Battle of Elkhorn Tavern and

Marmaduke's 1863 Missouri raid. When Shelby, Quantrill, and others fled to northern Texas, James was placed, at times, under Quantrill's command. The Ward letter states that James was in "this section" (Louisiana) in 1863-64 and part of 1865 in the Cavalry. (Ward erroneously assumes that Jason James was related to Jesse James.)

James declares that Lea was his battalion commander during the last two years of the war. James first became acquainted with Lea in 1863 and joined his command the following winter after his former unit fell apart.

James' description of Lea, who was six feet and four inches tall, was that he wore a dingy old Confederate uniform, a pair of six-shooters buckled around his waist, riding 'George,' his favorite old horse—which was a 'blood bay.' In James' estimation, Lea was born to command; had good judgment; and had a look of kindness, determination, and confidence on his face. He seldom made a mistake and never allowed his command to be surprised by the enemy. He was kind and just to his men. "When the good Lord made him He used the best material He had."

Louisiana State Troops. According to Padgitt, when the Confederate Governor of Louisiana, former General Henry W. Allen, heard that Kirby-Smith had refused to allow Quantrill's band to join the Confederate forces or even to stay in Louisiana, he invited them to come see him. Younger's account about reporting to Kirby-Smith contradicts Padgitt's assertion that Kirby-Smith ordered the guerrillas out of Louisiana. Padgitt writes that Allen asked the guerrillas to come under his direct command as members of the State Scouting Service. Quantrill rejected the offer, but many accepted. Padgitt claims that among those accepting the governor's invitation were the Lea brothers, the Younger brothers, and the James brothers, who elected Joe Lea as their captain. However, there seems to be some question about whether or not Frank and Jesse James were there—they probably were with Bill Anderson's guerrillas. Padgitt quotes documents that indicate J.C. Lea entered the service of the Louisiana State Troops on January 25, 1864. At the same time, the company was apparently still on the rosters of the Confederate States Army.

The home base of Lea's unit was Carroll Parish, some twenty-five miles west of Vicksburg on the Louisiana side of the Mississippi. This was right in the middle of the Cotton Belt, but much of the land was flooded because the levees had been neglected during the war. Lea's guerrillas stayed in the two little towns of Delhi and Floyd and on plantations in the area, where they found the local population to be both friendly and helpful. Padgitt states that J. C. Lea, John Jarrette, and Frank and Jesse James stayed with a planter by the name of Dickson. The guerrillas were the masters and the heroes of the parish, and they were wined and dined every night by some local planter or other. "They kept the community in a state of perpetual excitement and enthusiasm," Padgitt writes. "They made sad havoc among wives and maidens." Lea's men rode fine horses, carried fine side arms, and dressed gaudily, as the public seemed to expect of guerrillas.

When the company went on a raid behind Union lines, they wore Federal uniforms. That was one reason why the guerrilla movement, in general, was able to survive so well—Union troops never could distinguish between friends and enemies. According to Padgitt, "... Lea and his raiders decoyed hundreds of Yankee soldiers into the jaws of death .... There was no quarter given and certain death if captured."

Lea's company engaged in actions on the Mississippi between the Arkansas state line and the Arkansas River, according to Jason James. James joined Lea's company, which at the time was in northern Louisiana, as a private and soon was elected orderly sergeant. In a skirmish in the spring of 1864, Lea's first lieutenant, Jim Stevenson, was killed, so James was elected to take his place. Lea then received a commission to recruit a battalion of cavalry, and he signed up three additional companies. Lea commanded this battalion of guerrillas until the war ended, so James took over as captain of Lea's previous company. Lea and James became close friends, spending their off-duty times together, and "... often shared the same blanket."

James writes that in their service along the Mississippi River in 1864-1865, they were active at all times, winter as well as summer;

and they had many hard-fought skirmishes. The area's 100-inch annual rainfall made military operations quite difficult for horses as well as soldiers.

Atrocities at Lum's Plantation. One of the principal events in which James and Lea participated together took place in the spring of 1864. U.S. General James B. McPherson commanded the Union troops at Vicksburg. A road started opposite Vicksburg in Madison Parish and led to northern Louisiana, Arkansas, and Texas. McPherson was known as a kind and humane man, James avers, and gave permits for local women to travel this road to and from Vicksburg to buy goods. Manufactured goods also were carried along the route, and Confederate soldiers paroled by McPherson traveled it.

About fifteen miles west of the river was Lum's Plantation, on Willow Bayou (James gives the name as 'Walnut Bayou' in another rendition). Where the road crossed the bayou, two companies of Negro infantry totaling about a hundred men were stationed. Northern men who had taken over the plantation to produce cotton had hired them. The black militia waylaid and robbed women and merchants who passed by. More significantly, when the paroled Confederates came along the road, the soldiers killed them and threw their bodies under the bridge.

Captain Lea heard of these atrocities and sent Jason James with about thirty men dressed in Union uniforms to investigate. This advance guard rode into the log-cabin camp of the black militia, where they were served breakfast. Some of the militiamen took James' company to the bridge and showed them the Confederate bodies, some of which had been partially devoured by hogs. James sent this information to Lea, who hastened to the plantation with the rest of James' company and Captain Cy Hedrick's company. Lea (James refers to him as 'Major Lea' at this point) attacked the two companies of black militia. James writes that it was a short fight and a clean job. All of the one hundred militiamen were killed; none escaped. Lea suffered no casualties. The Confederates captured some weapons and ammunition from the militia.

Meanwhile, U.S. General McPherson had heard of the atrocities and determined immediately to put an end to "... this outrageous act of these savages." His honor was at stake because the militia were killing Confederate prisoners that were supposedly under McPherson's protection. He sent a white cavalry detachment to kill the militiamen, according to James, but Lea arrived first. When they were still three-quarters of a mile away, the Union men heard the noise of Lea's attack. According to James, the Union soldiers returned to their camp and told McPherson that the Confederates had done their job for them.

Shelby Foils Steele at Shreveport. Another campaign in the spring of 1864 certainly involved J.C. Lea, but it is not known whether it was before or after the incident at Lum's Plantation. In early April, U.S. General Frederick Steele marched south out of Little Rock in order to help U.S. General Nathaniel P. Banks capture Shreveport. CSA General Marmaduke was in charge of the resistance, and he ordered Shelby to annoy and retard Steele. Shelby had much success in several skirmishes with Steele, capturing several hundred Federals at one point. Near Camden, Arkansas, Shelby's men made a remarkable ride of ten miles in fifty-five minutes to aid troops under the command of an officer named Cabell. According to Edwards:

> ... the charge became like the march of a hurricane. Lines went down before it at a breath, and entire regiments surrendered without a shot. Captain Lea, with one company, took two hundred prisoners, and in marching them to the rear had scarcely men enough to form a respectable guard ... .

There was more fighting, but ultimately Steele's plan to take Shreveport was foiled.

Swapping a Yankee for a Rebel. Another incident related by James took place around October 1864, opposite Vicksburg on Bayou Roundaway. James was scouting there and came to the residence of Joshua James (no relation). Many ladies were present, and they prepared breakfast for James' company. Mrs. Joshua James related to Jason James that her non-combatant husband had been arrested by Federal troops and jailed in Vicksburg three months earlier. James later heard of a Yankee cotton-buyer in DeSoto, so he planned to capture him and swap him for Joshua James.

When Jason James told Captain Lea about his proposal, Lea took forty of his best men, dressed them in Federal uniforms, and went to DeSoto at night under a full moon. They rode through a black militia who were suspicious but let them pass. Lea apprehended the cotton buyer, and the next day he sent a flag of truce. The two prisoners were exchanged, and Joshua James was returned to his anxious wife.

Reporting to CSA General Nathan B. Forrest. In October 1864, Lea received orders to cross to the east side of the Mississippi and report to CSA General Nathan B. Forrest. Lea took part of the command and crossed the river at the head of Grand Lake, James states, by swimming the horses at the side of skiffs, with the men holding to the horses' halters. James came with the rest of the command later. The command then moved north into Bolivar County, Mississippi, where Lea and James reported to one Major Blackburn of General Forrest's command.

Union officers in Arkansas, Louisiana, Mississippi, and Tennessee were well-aware of Captain Lea, mentioning him by name in several of their dispatches, which are reproduced in the official history of the Civil War, *War of the Rebellion, Official Records of the Union and Confederate Armies, Series I,* Vol. XLI, Part IV (1893), pp. 128-29, 675, 901, and 951-52. On October 20, 1864, Brigadier General M. K. Lawler, commanding at Morganza, Louisiana, sent the following message to Colonel Edmund J. Davis at Morgan's Ferry:

> .... I am informed that a Captain Lee [sic], with 100 men of Quantrill's band, dressed in Federal uniforms, has made application to the officer commanding confederate forces at Simsport [Louisiana] to be permitted to cross to this side of the Atchafalaya. He refused permission and threatened to fire on them in case they should attempt to cross, as General Smith and the Confederate authorities regard them as outlaws ... .

The next day, Lawler sent the same message to General Headquarters at New Orleans, adding that Lea might attack farther downstream on the Mississippi.

Lea was not to be out-guessed by his enemy, however. Federal dispatches indicate that he was recognized during the trip to Bolivar County as related by Jason James. Lieutenant A.M. Jackson of the U.S. Signal Corps sent a message on November 25, 1864, to US Colonel C. T. Christensen at Headquarters, "... Captain Lee (guerrilla) was at Trinity, on Black River with 500 men, mostly deserters from Harrison's Brigade, and paroled prisoners ... ."

Major-General E.R.S. Canby, who had commanded U.S. forces in New Mexico Territory in 1861-62, was in command at New Orleans in December 1864. On the 21st, he sent an urgent message to General N.J.T. Dana at Memphis:

> Sir: I wish you as soon as possible to organize an expedition for the purpose of driving off the guerrilla and partisan bands under Harrison and Lee that now infest the upper parishes of Louisiana between the Washita and Mississippi Rivers, and desire that you will organize a force for that purpose as soon as possible ... .

Dana replied on December 28, that he could not do as Canby requested because his cavalry was absent. He allowed that once he got the 2,000 to 2,500 cavalry he needed, "... I can easily drive Harrison and Lee almost any where, but they will immediately return on our retirement ... ." Indeed, Lea typically would be chased away from one column's flank only to show up in the rear of another.

General Kirby-Smith at last took note of Lea's merit as a brave leader of soldiers. He made Lea a full colonel in a battlefield promotion early in 1865, according to Padgitt. A statement written many years later by Ella Lea Dow, confirms the promotion:

> He rose from lowest ranks to that of Colonel, by successive steps of private, corporal, orderly, sergeant, second lieutenant, and captain. Just before the close of the war he was made a Colonel by General Kirby Smith, but he has always been called 'Captain.'

End of the Civil War, 1865. When the officers and men of the Trans-Mississippi Department learned that CSA Generals

Robert E. Lee and Joseph E. Johnston had surrendered, ending the war in April 1865, it was as if an earthquake had hit them. They were shocked by what to them was a great, unexpected disaster. Many commands, except Shelby's, disbanded themselves and went home. J.C. Lea kept his army going for two more months and was discharged from the Louisiana State Scouting Service on June 2, 1865. Jason W. James surrendered his company to Colonel William H. Dickie on June 27. Lea and James apparently were protecting civilians from attacks by outlaws and keeping order in the communities in their areas during the final weeks of their military service. These activities were seen as having mostly local importance, and scant space is devoted to them in Confederate histories.

General J.O. Shelby had no intention of surrendering. Kirby-Smith surrendered on May 26, after which he and Shelby took about five hundred men to Mexico. Many other former Confederate generals—notably Sterling Price—and governors had the same idea, along with hundreds of Southern soldiers. According to Padgitt, Kirby-Smith and Shelby asked Captain Lea to go with them, but he had had enough of fighting. Padgitt states, "Unlike some of his contemporaries who had learned to live on plunder, he took off his side arms, and was never again known to wear a gun."

Many soldiers did not stay around for a formal surrender. Thinking the war was over, they simply went home. But some could not go home. In his memoirs, James states, "The feeling was so intense in Missouri that my brother and I could not go home to see our mother, like boys from other states. We were forced to remain out of the state or at least that part of it where we had formerly lived."

Unfortunately for history, General J.O. 'Jo.' Shelby's Sixth Missouri Cavalry Regiment's records were destroyed when the command separated at Corsicana, Texas, on June 2, 1865. On July 4, 1865, Shelby ceremoniously buried his undefeated battle flag in the middle of the Río Grande; and the last vestige of the Confederate Army disappeared as they crossed to the Mexican side. He offered his services to Emperor Maximilian, who declined at first but then accepted. Maximilian made 100,000 acres of land available to the Confederates, upon which they planted a colony named after the Empress Carlota. Shelby became a freighting contractor there and later started another colony at Tuxpan.

Even before the deposing of Maximilian by the French in 1867, the Confederate colonies in México were generally failures, according to W.C. Nunn in *Escape From Reconstruction* (1956). Hard times forced Shelby to return to Missouri, where he became a wheat farmer in Bates County. In 1893, President Grover Cleveland appointed him US Marshal. Shelby died on February 13, 1897.

Frank Lea shortly after the Civil War. *Courtesy: HCSNM, Photo No. 652.*

J.C. Lea shortly after the Civil War. Photo by Joslyn & Smith, Washington Gallery, Vicksburg and Jackson, Mississippi. *Courtesy: HCSNM, Photo No. 3326-B*

3

FROM RECONSTRUCTION TO NEW MEXICO TERRITORY

J.C. Lea During Reconstruction

Some of J.C. Lea's early activities following the Civil War are not clear. He may have gone to Tennessee and perhaps to Washington to secure a pardon.

In December 1863, President Abraham Lincoln announced his plan of Reconstruction, *i.e.,* for getting the former Confederate states back into their proper relationship to the Union. He offered a general amnesty to all white Southerners except Confederate government and military officials. When ten percent of the number of 1860 voters in a state had taken an oath pledging loyalty to the United States and accepting the abolition of slavery, that state could proceed to establish a state government. Tennessee, Arkansas, and Louisiana met the requirements by the end of 1864. Congress refused to recognize Lincoln's plan and proceeded to legislate its own radical plan. Unfortunately, Lincoln was assassinated on April 14, 1865.

Vice-President Andrew Johnson succeeded to the presidency. He was a former governor and senator from Tennessee—a Southerner and former slave owner. Johnson came up with his own 'Restoration' plan, which was a compromise between Lincoln's and Congress' plans. In the summer of 1865, while Congress was not in session, he set his plan in motion. The plan offered amnesty to Southerners who would take an oath of loyalty. Large landowners and high Confederate government and military officials would have to apply to the president to secure an individual pardon. Congress instituted its radical plan later.

The president pardoned J.C. Lea in September 1865. Ella Lea Dow, Lea's daughter, wrote in a letter that "... his uncle, Mr.

Thomas Callaway, through his personal friendship with President Andrew Johnson, secured for Capt. Joseph Lea an immediate pardon." Callaway lived in Cleveland, Tennessee, J.C.'s birthplace. Aparently, Callaway—a Union man—convinced the president that Lea was still a fellow Tennessean, although he had been out of that state for sixteen years. It is not clear whether Lea had to go to Washington, or even if Callaway did. On the back of the large (two pages folded over, 18x23 inches) pardon document is a notation in pencil: "Send to J.C. Lea, care of Thomas H. Callaway, Cleveland, Tenn." This document is in the J.C. Lea Collection in the archives of the Historical Center for Southeast New Mexico, placed there by his great-granddaughter, Lea Rohr Fraser. The text of the pardon is given on the following page.

"And from that moment on," Padgitt writes, "he dedicated his life to the support of law and order and good will toward his fellow man."

J.C. Lea was Jason James' best man and Frank Lea was an attendant at James' wedding to Mary 'Mollie' Henderson of Carroll Parish, Louisiana, on April 17, 1866, according to James in a talk to the Daughters of the Confederacy in Roswell in April 1909. On April 18, 1867, J.C. Lea married Ellen Burbridge, the widow of Douglas Burbridge, according to Verna Reed of Carlsbad, New Mexico. Ella Lea Dow writes that she believed her father and Ellen were married in New Orleans, but Ms. Reed's source indicates that they were married in Warren County, Mississippi.

Another issue concerning Lea's whereabouts after the Civil War involves his connections with Cole Younger. Carl W. Breihan, in his book, *Younger Brothers: Cole, James, Bob, John* (1961), states that Younger was in California as a Confederate recruiter at the time the Civil War ended. According to an article in Jerry J. Gaddy's book, *Dust to Dust* (1977), all of the Younger brothers—Cole, Jim, and Bob—were back on the family homestead near Harrisonville, Missouri, in 1866; but they were persecuted a lot because of their Confederate and Quantrill connections. Exactly what happened over the next two or three years is obscure, but it is clear that Cole went back to Louisiana for a while to escape those who pursued him because of his wartime actions.

ANDREW JOHNSON

PRESIDENT OF THE UNITED STATES OF AMERICA
TO ALL TO WHOM THESE PRESENTS SHALL COME, GREETING:

Whereas, *J. C. Lea, of Cleveland, Tennessee, by taking part in the late rebellion against the Government of the United States, has made himself liable to heavy pains and penalties:*

And whereas, *the circumstances of his case render him a proper object of Executive clemency:*

Now, therefore, be it known, *that I,* ANDREW JOHNSON, *President of the United States of America, in consideration of the premises, divers other good and sufficient reasons me thereunto moving, do hereby grant to the said J. C. Lea, a full pardon and amnesty for all offences by him committed, arising from participation, direct or implied, in the said rebellion, conditioned as follows:*

1st. This pardon to be of no effect until the said J. C. Lea shall take the oath prescribed in the Proclamation of the President, dated May 29th, 1865.

2d. To be void and of no effect if the said J. C. Lea shall hereafter, at any time, acquire any property whatever in slaves, or make use of slave labor.

3d. That the said J. C. Lea first pay all costs which may have accrued in any proceedings instituted or pending against his person or property, before the date of the acceptance of this warrant.

4th. That the said J. C. Lea shall not, by virtue of this warrant, claim any property or the proceeds of any property that has been sold by the order, judgment, or decree of a court under the confiscation laws of the United States.

5th. That the said J. C. Lea shall notify the Secretary of State, in writing, that he has received and accepted the foregoing pardon.

In testimony whereof, *I have hereunto signed my name and caused the Seal of the United States to be affixed.*

Done at the City of Washington this Seventh day of September, A. D. 1865, and of the Independence of the United States the Ninetieth.

By the President: *Andrew Johnson*

[SEAL] Will. H. Seward
Secretary of State.

The Youngers maintained they wanted nothing but peace, but their detractors claimed that they had become so accustomed to bloodshed and the excitement of warfare that they could not stop it. The three Younger brothers eventually banded together with Jesse and Frank James to enter the bank- and train-robbing business. In his autobiography, *The Story of Cole Younger* (1903), Younger attempts to refute many of the charges against him and his brothers. Their first job, it was said, was the bank at Liberty, Missouri, on February 13, 1866, which was the first bank ever robbed by a band of organized outlaws. The robbers took $58,000 and killed a bank guard. Cole writes that at that time he was visiting his great-uncle in Howard County, Missouri.

A bank in Lexington, Missouri, was robbed on October 30, 1866. Another robbery at Savannah, Missouri, was attempted on March 2, 1867, but failed. A third bank robbery took place on May 23, 1867, at Richmond, Missouri. Younger's defense of these crimes provides some clues to the location of J.C. Lea, if true.

Younger writes that Lea was on the "Widow Amos' farm" on Fortune Fork in Tensas Parish, Louisiana; and Younger went there to stay with him. After Younger had been there with Lea for three months, they rented the Bass Plantation, three miles below Lake Providence in Carroll Parish. Younger claims he stayed there until the fall of 1867, living in the same house with J.C. and Frank Lea, as well as their older brother, Thomas Callaway 'Tom' Lea.

Younger seems to have parted company with the Leas in the fall of 1867. The robbery of the bank at Russellville, Kentucky, on March 12, 1868, was the first to be publicly attributed to the Younger and the James brothers. Although Cole Younger denies it, other accounts maintain they continued robbing, together or separately, until the Younger brothers were captured following their robbery of the Northfield, Minnesota, bank on September 7, 1876. They were sentenced to life imprisonment. Bob Younger died in prison in 1889, and Jim and Cole were paroled in 1901. Jim took his own life the following year. Cole became a performer and lecturer. In 1913, he got religion and became an active church worker until his death in March 1916.

Back to Lee's Summit

When William B. Howard, the founder of Lee's Summit, returned to Missouri in 1865, he likely felt that it would not be prudent to continue with his plans to name his town after Robert E. Lee, the great Confederate general. He named it "Strother," after his wife's maiden name when he laid out the town in October 1865 and secured a post office later that year. According to local historian Donald Hale, bills of sale for lots stated that Strother was previously known as Lee's Summit. Further, it was still being called by the earlier name, as evidenced by an item in the *St. Louis Democrat* reporting that when the train stopped at Lee's Summit on December 26, 1865, a lively dance was underway. Howard decided to return to his original plan. In November 1868, Hale states, the town's residents petitioned the county government to change the name from Strother to Lee's Summit.

Howard secured a boxcar from the railroad to serve as a depot. If the town fathers intended to name it after Dr. Lea, someone misspelled the name when it was painted on the boxcar. The spelling was never changed. However, Howard's grandson, William T. Howard, maintained that his grandfather intended to name it after Robert E. Lee all along, according to Hale, and not after Dr. Pleasant Lea. At the turn of the 21st century, Lee's Summit is a thriving suburb of Kansas City. A park, an elementary school, and a junior high school all carry on the name of Pleasant Lea.

The 1902 yearbook for New Mexico Military Institute states that Lea secured a twenty-five percent interest in a contract to rebuild the Georgia and Central Railroad, which U.S. General William T. Sherman had wrecked in 1864. He next engaged in cotton planting in Louisiana from 1870 till 1872, but his health failed. At some point probably around 1870-71, J.C. and Ellen Lea apparently moved back to his home area or at least spent some time there. By that time the town was officially known as Lee's Summit. Ellen died there in 1871 and is buried there in the Lee's Summit Cemetery. J.C.'s parents are also buried in that cemetery. Originally, they were buried at their farm. In 1900, a descendant visited the place and discovered the current owner had a hog pen

Major William W. Wildy, 1809-1881, Yazoo County, Mississippi. Wildy was J.C. Lea's father-in-law, and he played a major role in the development of the Roswell area.
Courtesy: Lawrence W. Long, III, Yazoo City, Mississippi

gravesites. She immediately took steps to have the bones exhumed, cleaned, and re-buried in the city cemetery.

Following his wife's death, Lea traveled to the drier climate of Colorado until his health was restored. He then moved to Mississippi, probably Yazoo County. The dates given in the various sources do not agree, but apparently Lea's move to Mississippi was about 1873. He became a livestock trader there and in Louisiana and continued in that business until after he remarried and moved again to the West in 1875.

The Wildys of Yazoo County

In Yazoo County, Mississippi, J.C. Lea courted Miss Sallie Edith Wildy (pronounced WILL-dy), a cousin of Lea's first wife. She was a daughter of William W. Wildy of Satartia, Mississippi, a distinguished cotton planter.

William W. Wildy was a native of Northumberland County, Virginia, where he was born on October 9, 1809, one of three sons and four daughters born to William and Sarah Wildy. Census records suggest that William Wildy died sometime before 1818, and his widow moved to Rutherford County, Tennessee, in the 1820's and likely died there. Apparently, the Wildy family was part of a migration to take up lands in Yazoo County that had been noticed by General Andrew Jackson during the War of 1812. The plantations of the emigrants were in Yazoo County a few miles south and southwest from Satartia.

According to Harriet DeCell and JoAnne Prichard in *YAZOO: Its Legends and Legacies* (1976), there were three active communities in southwestern Yazoo County by 1830. Primarily people who had at first moved into the Old Natchez District from Tennessee or Kentucky settled the communities. Some of them came directly from the Carolinas or Virginia.

The largest of the three communities was Satartia, the only one to retain its Indian name ('pumpkin patch'). In that section, the plantations, farms, and houses were bigger; and the planters there owned more slaves than in other parts of the county. The Satartia area was the only part of the county where the typical ante-bellum plantation evolved. Satartia was platted about 1826 and incorporated in 1833. It was the center of activity in the 1830s and 1840s for that part of the county. By 1857, the town had passed its peak and saw a period of decline; it was re-incorporated in 1875.

According to Verna Reed, a Wildy great-great-granddaughter, William W. Wildy is not listed in the census of 1830. He likely moved to the Satartia area in the 1830s. Wildy had a close connection with Albert F. and Sarah Keeble, although family research has yet to uncover the exact nature of that relationship. Deed records show that in 1836, the Keebles bought the Madley Plantation from William Ferriday. On June 8, 1838, the Keebles and William W. Wildy sold the same 2,100-acre plantation back to Ferriday for $110,000. They also sold to Ferriday over eighty slaves, some livestock, and other "personal property" for another $127,904. What portion, if any, of this wealth came from the Wildy family is unknown, as is the nature of William W. Wildy's involvement in these transactions. Dr. C. Wildy Bonney, Wildy's grandson of Satartia, told a researcher that Wildy was first a merchant in Marion County and then a planter in Yazoo County.

W.W. Wildy Married Laurena Hall Childress. Wildy married Mathilda Laurena Hall Childress, a widow with two young sons, probably by 1840 since he is listed in the 1840 census as a head-of-household; and their first child was born in 1842. Unfortunately, neither Yazoo County nor the Mississippi State Archives has a record of marriages from that early period. Laurena was the daughter of Samuel S. Hall and Matilda Garrett Hall, born in Nashville, Tennessee, on August 3, 1819. Hall had been a staff officer for General Jackson at the Battle of New Orleans in 1814-15, and he was among those who later settled in Yazoo County. Laurena was first married to Judge Elijah Robertson Childress of Benton, Arkansas; and they had two sons: William and Samuel. After Judge Childress died, Laurena returned with her two young sons to Satartia and later married Wildy.

William and Laurena Wildy had several children, all born in Yazoo County: Mattie, born in 1842; Harry, 1846; Sallie Edith, born November 3, 1849; Ada, 1851; John, 1853; Ernest, 1856; and Ella,

1861. One source lists another child, Willie, who died young; but the child's identification is not certain. Also, according to Mrs. Reed, there is some evidence that the Wildys may have had twin girls who either were stillborn or died in infancy.

Wildy Plantation. Yazoo County records indicate that Wildy was the original owner of land southwest of Satartia (Township 9 N, Range 4 W). According to Lawrence W. Long, III, of Yazoo City, who is a descendant of Wildy's step-son William Childress, the lands of the Wildys and related families—Bonney, Hall, Childress —were three to five miles from Satartia, with some near Phoenix which is less than ten miles from Satartia. The number of acres in Wildy's plantation is unknown, but the 1850 census shows that he owned real estate worth $8,000 and had fifty-two slaves. That placed him in the influential social class known as the 'Planter Aristocracy,' that group of wealthy planters who exercised more than their fair share of influence and power in the society, the economy, and the government of the South. Only about forty planters in Yazoo County owned more slaves than did Wildy, but some six hundred owned fewer. The 1860 census indicates that his real estate had grown to $30,000; and he had personal property (including slaves) valued at $54,620.

Wildy was the quintessential Southern planter. An insider's look at life on his plantation is provided in an article quoted by DeCell & Prichard, written by a teacher from Battle Creek, Michigan, who went south in search of a position. A. DuPuy Van Buren wrote "Jottings of a Sojourn in the South" in 1859. He was surprised to learn that almost everyone, including girls and women, traveled on horseback instead of in carriages. He writes that the Wildy plantation was in the valley, and the old plantation house was located there on a knoll overlooking the entire plantation. It formerly was used by the family but in 1859 was the residence of the overseer. It was a large log house with a low porch. In this house, Van Buren had his first meal with Wildy and spent the night. The log cabin slave 'quarters' were located nearby.

The next morning, young Harry Wildy, thirteen, came to fetch Van Buren to the 'Ridge House,' the family house in the uplands about two miles from the old house. Van Buren was very impressed by the long strings of Spanish moss hanging down from the oak and hickory trees. Upon reaching the Ridge House, Van Buren "... met a cordial reception from Mrs. W., a lady of true Southern frankness—of a generous and spirited nature, and whose countenance expresses much of the feeling of her heart."

All the Wildy daughters and their teacher were just leaving on their "palfreys" for church. Van Buren noticed that the single women were called 'Miss Bessie,' 'Miss Sallie,' etc., while married women were addressed as 'Mistress Wildy,' 'Mistress Childress,' etc. He comments on Southern manners and concludes, "One might say that every-thing is different here from Northern life." The Southern staff of life was cornbread, and the sweet potato was also an important item on the menu. Clocks, Van Buren avers, were hard to find: "The planter takes his time eating; he takes no note of time."

Wildy in the Civil War. How the Civil War affected Wildy is not clear. Although the Confederacy eventually drafted men up to age fifty, Wildy was already fifty-one when the war started. One of his sons, Harry, served in the Southern army; and Wildy himself enlisted in a local militia unit recruited to serve as military scouts and probably to protect the home front as well. Records in the Mississippi State Archives show that Wildy enlisted at Yazoo City on April 16, 1862, as a private in Captain H. C. Tyler's company of Captain Gartley's Company of Cavalry, Mississippi Volunteers, known as the 'Yazoo Rangers.'

According to information provided via the Internet by Robert Crook, the Yazoo Rangers were Company 'D' of the Second Battalion of Mississippi Cavalry (Montgomery's) who functioned as independent scouts on the Mississippi and Yazoo Rivers. They were frequently under the command of Wirt Adams' First Mississippi Cavalry.

The unit was ordered into active service on May 14, 1862, at Liverpool; but two muster rolls in the records indicate that Wildy did not go on active duty with the outfit. Remarks on the muster roll for June 14 to November 1, 1862, are "Exempt by Age. No Pay." The roll of December 10, 1862, to February 28, 1863,

Sallie Edith Wildy at the age of twenty,
Yazoo County, Mississippi.
HCSNM

states, "Absent with leave. No pay." This suggests that Wildy's enlistment was primarily a show of support and loyalty for the Confederacy and the local troops. Crook states that the Yazoo Rangers many have participated, together with the "11th/17th AR Mounted," in April 1864 in the capture of the Union tinclad *USS Petrel* on the Yazoo River. Crook's great-great-grandfather, Isaac H. Hunter, lost an arm in that incident.

William W. Wildy was known as 'Major' for much of his life, but no explanation for that title has surfaced. Van Buren refers to him as 'Major' in his 1859 article, written before the war. Wildy obviously did not achieve the rank of major in the Civil War. Neither the National Archives nor the Mississippi State Archives has any record of a military career for Wildy during the Mexican War or any other time prior to the Civil War. No mention of how he came by the title of Major is made in any of the extant family papers, and no known descendant has knowledge of the subject. This leaves one to speculate that it was an honorary title conferred upon him somewhere along the way. Perhaps it was a nickname, or a self-assumed title in keeping with his status as a planter aristocrat.

To what extent Wildy may have lost his land or fortune in the war is not clear. He apparently kept his land and was able to rebuild his fortune in the years following the war. The 1870 census shows his real estate valued at $20,000; and he had personal property worth $330.

Sallie Edith Wildy Married Joseph C. Lea. It was in this setting of privilege and polite Southern society that Joseph C. Lea found Sallie Edith Wildy. They were married on February 3, 1875; Sallie was twenty-six years of age and Joe was thirty-three. "My mother was a woman indulged and accustomed to every luxury," their daughter Ella told Georgia Redfield in 1936. "My father insisted that she remain in her home until he could prepare a safe and comfortable place for her ... ."

New Mexico Territory

Lea soon left for New Mexico Territory, while Sallie stayed behind with her parents. He hoped that ranching on the frontier would afford a better prospect for his future than did farming. Lea took over the old Cooper Ranch on the Red River in Colfax County and started raising sheep. This Red River is not to be confused with that of Texas and Oklahoma; it runs north from Wheeler Peak a short distance, then turns west for another short span and empties into the Río Grande. Lea's ranch was near the gold mining area around Elizabethtown in the high country north of Wheeler Peak, New Mexico's highest mountain. He likely leased the Cooper Ranch and also used public lands for grazing. There is no evidence in the deed records of Colfax, Mora, or Taos counties that he ever purchased any land there.

According to George F. Gerling in "Lea, Wildy, and Dow Families," *Roundup on the Pecos* (1978), Sallie wrote Lea that she would choose the hardships of living on the frontier with him rather than to remain in the comfort of her parents' home. "I want to be with you in New Mexico," she wrote, "even if there is nothing but the stars over my head." She joined him in Colfax County in early 1876. Harry Wildy Lea, their first child, was born on the Cooper Ranch on April 23, 1877. In her notes in the Lea papers, Ella Lea Dow states that her brother was born on the Cooper Ranch in Colfax County near Watrous. This does not seem to be accurate, because Watrous is some distance removed from the Red River area and from Colfax County. Another possible explanation is that there was another unit of the Cooper Ranch where Mrs. Lea went to deliver her child.

Sallie obviously named her son after her brother, who may have died shortly before this. According to Verna Reed, one family tradition holds that Harry Wildy was killed in the Civil War as a teenage drummer-boy. However, he is listed with the family in the 1870 census and inherited land when his mother died in 1872. Another family story says that he was an engineer on a railroad construction project in Mexico where Yaqui Indians killed him in the 1870s. He is not listed as an heir when his father died in 1881.

James D. Shinkle, in *Fifty Years of Roswell History, 1867-1917* (1964), states that the area Lea had chosen for his ranch was not ideal, not only because of the harsh winters but also because of the excitement and lawlessness of the miners' frontier. The Leas were open to new possibilities.

J.C. Lea about the time he moved to New Mexico Territory.
Photo by Schluter, Kansas City.
Courtesy: HCSNM

4

VAN C. SMITH AND THE FOUNDING OF ROSWELL

Early Activity

Although he was not the founder of the town, Captain Joseph C. Lea came to be called 'The Father of Roswell' because of his accomplishments in developing the region. Others had started the town several years earlier, and a limited amount of development had gone on in the Roswell area before the Leas arrived.

No record seems to exist which can pinpoint the time of the founding of Roswell. Lucius Dills wrote a booklet on the history of Roswell in 1933, *Roswell: Some Facts and Observations Relative to Its Settlement and Early Growth,* in which he summed up a problem faced by historians: "The exact, or even closely approximate, date for the initial settlement of what is now Roswell has been lost, as all of the participants have gone and no records have been preserved."

Conventional accounts attribute the first two buildings to Van C. Smith as the founder of Roswell. However, there are some other names in the mix that pre-date Smith. The first building in Roswell was a 15x15-feet adobe room, apparently used as a trading post. From photographs and insurance maps, it can be determined that this trading post was located at what later became 414 or 416 North Main. Depending upon whether the building was built at the front or the back of the lot, the west wall was either right on, or about twenty-five feet east of, the line between sections thirty-two to the west and thirty-three to the east (Township 10 South, Range 24 East). The east side of this one-room adobe was either twenty-one or forty-six feet west of the frontage line of the sidewalk as it was later developed.

Hiram Fellows surveyed the Second Standard Parallel in 1866, which is the south side of present Second Street east of Sunset

Avenue in Roswell. The following year, Fellows sectionized many of the townships in the area. Those who put up Roswell's first buildings were either unaware of the section lines or chose to ignore them.

Will Robinson, editor of the *Roswell Register-Tribune,* interviewed G. W. Hartman in 1910; and in an article in the March 15 issue, Robinson claimed that Hartman:

> ... was the original man in the Good Old Town. He came here just at the close of the Civil War and in 1867 built the old adobe house so many years occupied by the late Capt. J. C. Lea .... It was a matter of much labor. The glass was freighted from Ft. Sumner, where it was bought of the government. The first use of the building was for a primitive hotel .... Hartman was a boy then .... He sold his holdings in '69 to Van C. Smith and afterwards ran a store at the Block crossing .... .

It is doubtful that the place belonged to Hartman. He may have helped Smith build on to the original room, but his recollection of the year is probably incorrect.

About twelve years earlier, Dills, who was the first editor of the *Roswell Record*, published an interview with Aaron Wilburn in the April 22, 1898, issue. Dills states that Wilburn was "... a resident of Roswell in 1868 when one house constituted the town .... He and J.W. Hulbert started a store here; and shortly after, they sold out to Van C. Smith and his brother, Frank Wilburn .... ." It seems possible that 'J.W. Hulbert' was actually G.W. Hartman, and Wilburn just remembered the name incorrectly thirty years later. There is no known record of Hulbert, but there are several proofs of Hartman's activities in the area. The information about Wilburn, Hartman, and 'Hulbert' cannot be reconciled with other accounts which do not mention these men.

Van C. Smith

An unsigned article in the *Roswell Record* on April 8, 1892, titled "Roswell: Something of Its Past History," is one of the earliest attempts to systematically record the history of the town.

Artist K. Gunnor Petersen's conception of Van C. Smith, based on historical descriptions. Smith was the founder and first postmaster of Roswell.
Courtesy: Stu Pritchard

Drawing of Ash Upson by K. Gunnor Peterson, based on a portrait. Upson was an employee of both Van C. Smith and J.C. Lea, and an early postmaster of Roswell. His writings are an important source on the early history of the Pecos Valley.
Courtesy: Stu Pritchard

Based on the writing style, the article is usually attributed to Marshall Ashmun 'Ash' Upson, friend and employee of Van C. Smith and postmaster of Roswell, 1877-79. Editor Dills, who was to write his booklet on Roswell history some forty years later, may well have written the article. The article states:

> Early in 1870, Van C. Smith purchased from Jas. Patterson, all his interests in this immediate section: house, goods, and a considerable stock of horses and cattle. Just previous to this deal, Patterson had built an adobe house, one room 15x15 ... .

Patterson was one of the earliest cattlemen in the Pecos Valley, and he had a trading post in his little adobe building for the convenience of drovers using the Goodnight-Loving Cattle Trail. Smith made a squatter's homestead claim on the land where the buildings were located and nudged rancher A.M. Hudson off the grazing land Hudson had been using. All of this happened about the time Lincoln County was being created and organized.

Smith, the 1892 article further states, immediately upon buying Patterson's holdings:

> ... sent to Lincoln for adobe-makers and layers, and in an incredibly short time had three rooms added to the original one [*i.e.,* Patterson's 15x15 room]; sent to Fort Stanton for lumber and carpenters; raised the roofs; made the building one and one-half stories high, and placed the first shingle roof ever seen in the Valley.

Dills describes the hotel as being "... situated about the middle of the block, facing east, and was a four-roomed adobe, with a dormer-windowed shingle roof enclosing an attic for sleeping quarters for the guests." Pictures and insurance maps confirm that Smith's building was located at what was to become 414-16 North Main (astride the line of lots 4 and 5 in block 4 of the original townsite of Roswell). The west wall of the building was virtually on the section line. The earliest Sanborn insurance maps for Roswell (1900, 1905, 1909) show that the building was some forty feet from east to west, and about thirty-one feet wide. The east side of it was about twenty-one feet west of the frontage line of the store buildings that were built on the block in the early twentieth century.

During the building project, Smith entered into a partnership with Frank Wilburn (most sources claim, apparently erroneously, that the partner was Aaron Wilburn, Frank's brother) to engage in the mercantile business. The 1892 article states "... immediately after the organization [of Smith & Wilburn] was completed, the firm commenced the erection of the store building ... ." The article does not describe the store, but Upson, who later worked in the store, described it in a letter written to relatives on August 30, 1876, and reproduced in Maurice Garland Fulton's *Roswell In Its Early Years* (1963):

> Van Smith put up two good buildings—adobe of course. One, a dwelling—one and one-half stories high—square—four rooms below and one above. The other larger—square—one-half for a store, and the back divided into two rooms and a half story above ... .

Actually neither building was square, and the store was larger than the house. The 1900 insurance map suggests that the store was about the same width as the house, but longer. It extended about forty-eight feet from front to back and crossed the section line by some six to eight feet. Dills says this store building was "... located substantially where the Chamber of Commerce now functions" [*i. e.,* 1933, 400 North Main].

The unidentified 1892 writer continues:

> Whilst these improvements were being completed, the firm had on the road, from the Missouri river, an immense stock of goods, and the hundreds of cattle trains made it a point to stop as long as practicable, at the best supplied outfitting establishment on the [Goodnight-Loving] trail.

Upson continues:

> … [Smith] built a blacksmith shop, stables, chicken house, two very large corrals—one for horses—and one for cattle [old-timers' accounts written in 1931 and 1935 say the corrals were 50 ft. west of the house]; he set out trees all about the houses, brought water from North Spring River by acequias in front and behind the houses; built three farm houses on the Hondo within a quarter to a half mile distant ... .

An article in the *Roswell Morning Dispatch* in March 1937, states that cottonwood trees were brought from Bosque Grande by John Riley, an employee of Smith, in 1869 and set out along present North Main Street. Those trees were the beginning of Roswell's reputation as a town of trees, and some of them lived until they were removed in 1937.

With their store and hotel, Smith & Wilburn offered an excellent assortment of comforts and pleasures for the weary cattle-drovers who frequented the place while their herds passed through the area on the Goodnight-Loving Trail. The 1892 article declares:

> The best whisky and cigars that money would buy, faro, monte, poker and other, short-card games allured many, and toothsome viands many others ... . There was no luxury that money would buy that could not be found at his table. When these crowds assembled, dinner was on the table from 12 to 3 o'clock if necessary, and at $1 per head, whilst the bar-keeper was kept busy supplying straight whisky at 25c per drink.

In a facetious account written while he was in jail in Las Vegas in 1878, John Chisum implies that Smith & Wilburn may have offered other services as well. Speaking of Santa Fe—not Roswell—Chisum allows that Smith was engaged in the "income business," which Chisum defined as "... living with a good looking Mexican woman who lives in town and has a room and a clean Bed & has an income from her costimers and she divides her income with her lover." Frederick Nolan in *THE LINCOLN COUNTY WAR: A Documentary History* (1992), takes this to mean that Roswell's first 'hotel' was the first of the bawdy houses which would be a part of the Roswell scene for the next hundred years, and Smith was the original Roswell pimp!

The land upon which Smith started his town was not yet open for filing when he put up his buildings, although he borrowed money from Thomas Catron on September 24, 1870, as if he had filed on it. When the land was opened for filing, Smith was the first one to file; he filed claims on March 4 and October 14, 1871. In legal terms, the land upon which the buildings of Roswell stood was in sections thirty-two and thirty-three of Township 10 South, Range 24 East.

There has been some confusion and various interpretations about exactly where the first buildings were located in terms of the streets that were laid out later. Upson's and Dills' accounts are corroborated by that of Joseph 'Joe' D. Lea. The *Roswell Register-Tribune* carried on October 11, 1906, an interview with Lea, the first publisher of the *Roswell Record.* Lea said that he arrived in Roswell in October 1879, when:

> ... there were only three houses here then and only one of those remains, the old adobe residence of Capt. Lea. The other two, the postoffice and store building, which stood right here ["north-west corner of Main and Fourth streets" - Shinkle], and the blacksmith shop ..., have long since been torn down ... .

Joe Lea was a boy of twelve at the time; his father was Captain J. C. Lea's brother, Frank.

The 1892 newspaper article does not state that the store building was made of adobe, although Ash Upson's letter to relatives in 1876 claims both structures were of adobe. Upson tells how he first came to the area to survey land for John Chisum and stayed. Since he also worked in the store in that building, he should have known what materials were used to build it. Virtually every other account written since that time says it was of the same material as the hotel/house.

In the earliest photograph of the hotel/house, which was taken about 1884, it is easy to see that the hotel/house was made of adobe. However, the only known picture showing both the store and the house, taken in 1883, plainly shows that the store building was NOT adobe but some type of building blocks—probably manufactured stone. No historical account mentions that the building was stone. Of course it is possible that the adobe walls were sided with stone sometime before 1883 by J.C. Lea; otherwise, Upson surely would have mentioned the stone in his accounts. When the store was razed in 1900, it was referred to as adobe.

The 1883 photo also reveals that the two buildings were about the same size and design, with the store being located south of the house. The original plat of Roswell in 1885 shows, and several historical accounts confirm, that the store was at or near the northwest corner of Fourth and Main Streets. One account states that the store was just north of that corner, and another claims the buildings were eighty feet apart. Projections based on the 1900 insurance map suggest that the buildings were more like 145 feet apart. In the 1883 photo, the store appears to be in line with the house in terms of the distance west of the street.

Joe D. Lea stated in the 1906 interview that when he arrived in Roswell in 1879, the blacksmith shop was "... over there [on the site of Forstad & Johnson's Grocery store. This is the present location of E. F. Hutton Co. 1966 – Shinkle]; (the store and blacksmith shop) have long since been torn down to make room for more modern business blocks." The blacksmith shop was on the southwest corner—more-or-less—of present Fourth and Main, and it too sat back from the present street in line with the store and the house. Further, when James Hampton built a store in the middle of the 300-block in 1882, it also lined up with the other three buildings, assuming the 1885 plat is accurate.

Smith's Racetrack. One contemporary, Lily Klasner in *My Girlhood Among Outlaws* (1972), writes that Smith's store was "... not popular in the country ... because its owner was a sporting man and his life too fast for the more circumspect citizens of the lower part of the county."

Smith was indeed a "sporting man," *i.e.,* a gambler. Soon after the buildings were finished (c. 1869-70), the 1892 article reports, "... He laid out two parallel half-mile race tracks, now [1892] plainly visible, from his store to the Rio Hondo, and built a fantastic judge's stand near the store ... ."

Smith went to the East and returned with racehorses and gamecocks; he also kept a bulldog and beagle hounds to increase his repertoire of gambling offerings. Games of chance were offered not only to locals and passers-by, Upson writes, but:

> ... his place was visited by dozens of sporting-men from Santa Fe, Las Vegas, Albuquerque, and even from the States. Horse-racing, cock-fighting, dog-fighting and badger-baiting furnished daily amusement, whilst card-playing continued, often, throughout the night ... .

Joe D. Lea said that when he arrived in Roswell in October 1879,

> ... there were only three houses here ... . There was a race track commencing about where Price's store is located [present location of Radio Station KBIM. 1966 – Shinkle] (308 North Main), and terminating at the Hondo, about where Neighbor Gayle afterwards erected his residence [412 S. Main]. There were many exciting pony races and foot races on that old track. Charlie Ballard and I were urchins about twelve years old and we would run foot races together for the amusement of the crowd of neighbors who would sometimes come to the Post Office ... .

Lea's account seems to imply that the racetrack was the same as later Main Street, as some have assumed.

In a talk to the Roswell Woman's Club in August 1923, Amelia Church said when she came to Roswell in 1880, "... there was still remains of the race track he [Van C. Smith] had built across the flat from a point about where the Court House now stands, to a point near the Missouri Avenue school." It seems logical to conclude that Mrs. Church meant "even with present Missouri Avenue School," since there was no such school in 1880. If the track were very near where the school was later built, the starting line would have been considerably west of Main and the direction of the track would have been much to the northeast.

Mrs. Church's description of the track refutes the idea that it was the same as Main Street. If it had been, the 1892 writer would surely have said so, rather than stating that the two tracks were "still plainly visible." From these accounts, it can be concluded that the finish line and judges' stand of the race track were near present Fourth and Main Streets; and the starting point was a half-mile south of present Fourth and west of present Main, which gave the track a north-northeast orientation from start to finish.

Smith's Background. While Van C. Smith is generally recognized as the founder of Roswell, what is known about him is rather scanty. Frederick Nolan published an article in *New Mexico Historical Review* in April 1997 under the title "Van C. Smith: 'A Very Companionable Gentleman,'" which is the first account of the details of Smith's life. Van Ness Cummings Smith was a third-generation New Englander, the son of Roswell and Harriet Smith. He was born in Ludlow, Windsor County, Vermont, on July 12, 1837. He arrived in the vicinity of Prescott, Arizona Territory, shortly after that territory was separated from New Mexico in March 1863. He became one of the founders of Prescott and helped build the governors' log mansion. Smith became the first person ever to hold the office of sheriff in Arizona when the governor appointed him sheriff of Yavapai County (Prescott), where he served from June 1864 till January 1865. Smith was later honored as one of the first ten American settlers of Arizona.

After many attempts at mining, gambling, and other ventures in Arizona, Nebraska, and Las Vegas, New Mexico, Smith came to the Pecos Valley about 1869 or 1870 and put up the hotel and store buildings. Ash Upson's letter from Roswell on August 30, 1876, to relatives in the East, described Smith's career:

> He was a constitutional gambler. He next built a cockpit and race track with judges stand, etc. His gambling friends would come 250 miles from Santa Fe and Las Vegas and spend a few weeks [;] horse racing, dog fighting, chicken fighting, poker, etc., was the order of the day ... . Van had named his place Roswell, this being the name of his father. He had succeeded in getting a post office established, and there was no reason why he should not have had thousands of cattle, horses, sheep, hogs, etc., roaming over miles and miles of inexhaustible pastures (in winter as in summer) except that he could not refrain from

gambling, nor stay away from the cities where he could indulge his passion.

Roswell Post Office

Smith's place was first called 'Río Hondo' because an Hispanic settlement a mile or so east on the Río Hondo went by that name. To distinguish his settlement from the Hispanic community, Smith placed a notice in the Santa Fe *Daily New Mexican* on May 25, 1872, that henceforth his place was to be called 'Roswell'—his father's name. That same year, Smith left Roswell in the hands of [Frank] Wilburn, according to Nolan, and went back to Las Vegas. Smith had petitioned and applied for a post office in 1872, which was approved by the federal government. It opened in the Smith & Wilburn store on August 20, 1873, and Smith was back in Roswell for the opening. It was officially 'Roswell,' and Smith was the first postmaster. The office was a station on a Star Route between Las Vegas and La Mesilla.

Van C. Smith's younger brother, George R. Smith, was appointed postmaster on February 11, 1874, and served for about three years. The post office then went through a series of postmasters in a short time. Van C. Smith's old friend from his mining activities in Arizona Territory, Follett G. Christie, had been operating the store for Smith for some time when he received the post office appointment on February 16, 1877. About that time, Marion Turner finagled Smith and Wilburn out of their land claim and took over the store, so Christie was out after a couple of months. Daniel Carroll was appointed on April 25, 1877, and served about six weeks; nothing else has been learned about him.

Smith's Later Life

At the end of 1875, Smith left Roswell, never to return. He had a casino in Santa Fe for a while. His luck took another nosedive around the first of 1877. His father, acting with power-of-attorney, mortgaged the Roswell property to Santa Fe Ring leader Thomas B. Catron (again). At that time, John Chisum learned Smith had not 'proved up' on his homestead claim and therefore did not have the title to the land. Chisum acted to keep the Santa Fe Ring from devouring Roswell by urging Marion Turner, a cattleman who lived at Seven Rivers, to file on the land under the Pre-Emption Act. Turner paid Catron $800 for what Smith & Wilburn owed on the buildings and took over the property.

Nolan states that Smith left Santa Fe about 1877 and resumed his prospecting. About June 1879, he began serving as a scout for the 6th Cavalry; later, he was in charge of some twenty-five Indian scouts out of Fort Huachuca, Arizona Territory, during the campaigns against the great Apache warrior, Victorio. In 1880, Smith was back in the mining business and also was a deputy sheriff in Pima County, Arizona Territory. In 1881, he served as a deputy and ex-officio tax collector in Tombstone County. From October 1881 through January 1882, Smith was the chief of scouts at Fort Cummings, New Mexico. After that, he engaged in prospecting and ranching for the rest of his life, most of the time in Mexico. Smith died on August 29, 1914, in the Arizona Pioneers Home in Prescott and is buried there.

Nolan maintains that history gave the title 'Father of Roswell' to the wrong man:

> ... [Van C. Smith is] arguably more entitled than Joseph C. Lea to be remembered as 'the father of Roswell.'... . Throughout his life he was a nearly great man, a visionary, and an enabler, who, had he not allowed the lure of the gambling tables or the prospect of striking it rich to divert him, could easily have become an eminent civic figure in either New Mexico or Arizona ... .

Nolan concludes that Smith was at least, in the words of his obituary, "... entitled to the honor of being called the pioneer of pioneers, and to such small comfort as he received at the hands of those whose way he had made safe and clear."

5

J.C. LEA AND THE EARLY DEVELOPMENT OF ROSWELL

Most accounts of the early days of Roswell state that in 1877, J.C. Lea's father-in-law, Major W. W. Wildy, journeyed to New Mexico Territory in search of investment opportunities. He was impressed with the Pecos Valley and the Roswell area as a possible location for agricultural development, as the story goes, so he bought 480 acres of land from Marion Turner. The property included the land and buildings—the adobe hotel and store—where Van C. Smith had founded Roswell some seven or eight years earlier.

The Chaves County deed records, however, do not support this scenario. It seems likely that Wildy took an option on the land but did not purchase it until the following year. One possibility is that Wildy was waiting for Turner to get his titles clear so he could sell the land to Wildy.

Wildy's enthusiasm for Roswell enticed Lea and his bride to pull up stakes in Colfax County and move to Roswell in search of ranchland. Since they had not bought any land, it did not take long for them to prepare for the trek to Southeast New Mexico. Their entourage consisted of their baby boy, several herders and some of their families, a good-sized flock of sheep, and a small herd of cattle.

Most sources, such as Carole Larson in *Forgotten Frontier: The Story of Southeastern New Mexico* (1993), give the date of the Leas' arrival in Roswell as the fall of 1877; but Maurice G. Fulton, in *Roswell in Its Early Years* (1963) claims they did not arrive until February 12, 1878. Fulton further avers that the Leas could not move into the Smith & Wilburn hotel when they first arrived, so they moved into a small adobe house that Van Smith had built on the banks of the Hondo, south of the store. The reason for the delay was that Turner had taken Heiskell Jones as a partner and they were operating the store as 'Turner & Jones.' Turner and Jones also continued to operate the hotel for a while longer. Jones' wife, Barbara—better known as 'Ma'am Jones of the Pecos'—ran the hotel as a boarding house for a brief period. Turner later sold several tracts of land and improvements to Major Wildy.

Lea filed on the land directly west of Wildy's (Section 32, Township 10 S, Range 24 E). Since the Smith & Wilburn store building was astride the section line, the land claimed by Lea was that upon which the west end (back) of the store rested.

The Leas immediately set about improving the store and establishing law and order in the vicinity. One of his contemporaries, J.M. Miller, later wrote that when Captain Lea took charge of things, "A change for the better certainly took place at once ... . Lea had no patience with lawlessness, especially mob violence ... ." In his book, *Fifty Years of Roswell History, 1867-1917* (1964), James D. Shinkle tells a story that was related to him by Captain Lea's cousin, J. Smith Lea. It illustrates the Captain's approach to keeping order in his town:

> Shortly after Captain Lea took over the property and store in Roswell, four or five of the Seven Rivers crowd rode in and attempted to "Hurrah" the place by dashing up in front of the store, yelling and firing their pistols, and trying to ride their horses on the store porch.
>
> Captain Lea soon stepped through the store door with a Winchester and confronted the troublemakers.
>
> He told them that as long as they cared to come to Roswell and conduct themselves properly they were welcome. If they could not do that to leave and not come back.
>
> This was a shock to these men who were accustomed to "taking over" the small settlements of the area. They looked about and saw a least two other men with Winchesters at the store windows. They quietly left town. The word must have got around, as the performance was not repeated.

Georgia B. Redfield, an historian of early Roswell, wrote in 1937,

With her own hands she [Sallie Lea] cleaned and white-washed the walls of the adobe hotel where they established their home ... Captain Lea, was of commanding appearance. He was a leader of men to whom the pioneers of Roswell were glad to concede leadership, for the more than a quarter of a century that he lived in their midst. He and his wife ... worked continuously for the good of the town, for its up-building and for the educational and cultural developments not only for Roswell, but for the county and for the Territory of New Mexico.

The earliest description of Roswell as it was in the early Lea years comes from Captain Lea's niece, Carrie Lea Wells, daughter of Lea's brother, Frank. The Frank Lea family first arrived in Roswell in 1879. Carrie wrote an article for the *Roswell Daily Record*, August 30, 1935, describing what they saw. The boundaries of the village at that time were the North Spring River on the north, the Río Hondo on the south, and on the east and west prairie starting about a block either side of what was to become Main Street.

The hotel built by Van C. Smith was the home of the J.C. Lea family, and it also served as a boarding house and inn for travelers, stagecoach drivers, and anyone who needed meals or lodging. Mrs. Wells writes:

It was built of adobe, plastered inside and out. There were four rooms below and one large room above with upper and lower porches on the east side. The large upstairs windows on the east made the bedroom very cool and comfortable. The thick adobe walls made a cool home in summer and warm in the winter.

There were two front doors and the windows were shuttered ... . The northeast room was the living room and had a large fireplace and mantle in the northwest corner. The other front room south of the living room was used as the dining room. A long table seated the family and the ever-welcome guests. There was a bed room in the northwest corner of the house with a fireplace and mantle in the northeast corner. On the southwest corner was the kitchen, with a built-in table on the west side. There was a stair-way between the kitchen and the bed room.

Captain J.C. Lea's home and boarding house, the first building in Roswell. Photo taken *c*. 1884-88, after the death of Sallie Lea. *l.-r.:* Jack Galbreath (cook), Milo Pierce, Rufe Dunnahoo, Mary Lea, Frank Lea, Helen Holloman, Pinkie Daniels or Becky Camp, George T. Davis, Martha Hood, Smith Lea, Essie Fountain, Lelia Fountain, Ella Lea Calfee, Ernest Edwards, Oregon Bill, Ada Wildy Edwards, Mr. White, photographer. Most of the people shown are relatives of J.C. or Sallie Lea, but they and their children are not present. *Courtesy: HCSNM, Photo No. 573-D*

> Many dances and other meetings took place there. The front porch and tall cotton-wood trees in the yard were a favorite place in which to spend a summer afternoon. From the yard we could see the ox-wagons coming over the north hill, bringing goods from the outside world ... .
>
> West of this house about fifty feet was another adobe building used for the dairy ... . A ditch of water from Spring River ran between the dairy and the house. Another large ditch of water ran from the river, north of the house, going through the pasture just east to Main street, and crossing the Hondo in a flume, and so on to the farms east of town.
>
> A wire fence enclosed the yard. The store and post-office were just south of the fence ... .

Ella Lea Dow told Georgia Redfield in 1936, "An old sweetheart of my mother's came to visit us here in our western home. He was shocked to find her mounted on a soap box whitewashing the walls of our four-room adobe house ... ."

Another article with no author identified but probably written by Carrie Lea Wells also, was carried in the October 7, 1931, issue of the *Roswell Morning Dispatch.* The article includes much that was repeated in the later article, but the writer also describes other aspects of the settlement not included in the 1935 article.

> Just west of the Lea home were corrals built of adobe with thick high walls that seemed to defy the cow that jumped over the moon, and many were the long horns that tried it. North of the corral was a large peach orchard which besides yielding fine fruit, provided the tools for the switching we [children] needed so often.
>
> Between the orchard and house was the chief attraction of all—the dairy. Built of strong adobe walls, it kept milk cool all day. The floor was hard earth, almost as hard as cement. Many of the houses there at that time had such floors. Sprinkled and swept every day and with sheep skin rugs scattered around made them most comfortable. There were shelves on both sides of the dairy walls to hold the shining milk cans. They milked twenty or thirty cows most of the time ... . It was a treat for we children to be there when the cream was skimmed into the churn, and we were allowed to eat our fill of the clabber.

> North of the orchard were the bunk houses for the cowboys. South of the store and post office was a blacksmith shop, or better, a general work shop. For besides shoeing the mustangs and repairing the stage coaches, carpenter work was done, even to make coffins .... As we were over two hundred miles from the railroad that was the best that could be done. All of our merchandise had to be brought in by oxen yoked to large wagons.

Amelia Bolton (Church) was an early visitor to Roswell. In 1923, she stated,

> The first time I saw Roswell was in August 1880. At that time, I lived in White Oaks ... . We had heard so much of the wonderful fishing there was in the rivers around Roswell that we made up a party of friends and came down to fish and camp. We camped ... on the bank of Spring River, about where Haynes Park [Cahoon Park] is now. We had only been in camp a short time when Capt. Lea sighted us and drove over to see who we were. He was very pleasant, chatted with us awhile, and with the great-hearted hospitality for which he was famed, invited, even insisted on our coming to his house instead of camping. This, of course, we could not do, there were too many of us, but we did accept his invitation to lunch the next day.
>
> Just as he was about to depart, a six-yoke team of oxen, hauling a load of lumber and driven by the tallest man I ever saw, passed along. I remarked about his height. Capt. Lea said, "That is Pat Garrett; he will be our next sheriff." Garrett was elected the following November and afterwards cleaned the county of many its bad men, including "Billy the Kid."
>
> We remained in the vicinity of Roswell ten days. There were then four buildings here: Capt. Lea's home, Mrs. Ella Calfee's home, a store and post-office and a blacksmith shop ... .

Postmaster Ash Upson

Van C. Smith was again appointed postmaster in June 1877, but by then he had been gone about two years. According to New Mexico postal historian Jim White of Farmington, Smith probably accepted the appointment to prevent the closure of the post office, although he did not actually serve. The job then went to Ash Upson, who kept it until February 1879.

Upson moved the Roswell Post Office out of Roswell for a brief period during his tenure as postmaster, most likely at the height of the violence of the Lincoln County War in 1878. He 'opened mail' at his home, located one and a half miles east of the store. Evidence suggests that Captain Lea closed his store during the time that Upson had the post office at his homestead. William A. Keleher, in *The Fabulous Frontier* (1982), quotes a letter that Upson wrote to his sister on September 25, 1878, which implies that the store had been closed:

> Capt. Lea, who owns the buildings comprising Roswell, will put in a stock of goods in a very fine storeroom, next to my office in a few weeks—soon as he is sure the bandetti are effectually dispersed, and I am to take charge of the store, as Capt. nor any of his people talk Spanish, and if they did, would not have time to attend to it, and would not anyhow. This will pay board, if no more. The Postoffice pays very little but it is not much trouble.

This letter also implies that before Lea re-opened the store in the fall of 1878, Upson had already moved the post office back to the store building.

According to Keleher, Upson quarreled with Lea while clerking in Lea's store. A woman came in and asked for a bottle of 'Hostetter's Bitters,' a favorite tonic of the times, which had a plentiful allowance of alcohol in it. By chance Lea was present and saw that Upson could not find a bottle of the patent medicine for the customer. Lea was confident that a sizable supply of the bitters was in stock. Finally, Upson, who was known to have an alcohol problem, confessed that the reason there was no bitters in stock was that he had drunk it all himself. "How in hell was I to know that some damned fool woman would come in here, wanting a bottle of bitters," he raged. Taking umbrage at Lea's attitude, Upson then resigned his position, date unknown. Lea was appointed postmaster on February 24, 1879, and Upson probably continued working in the store after that. Lea served until April23, 1884.

Roswell in 1883, northwest from 4th Street. This is the only known picture of both the store (left) and the hotel (center) built by Van C. Smith, *c.* 1870. *Courtesy: HCSNM, Photo No. 376*

Wildy Family

The conventional version of the early history of Roswell is that Major W.W. Wildy gave his land to daughter Sallie Lea on August 1, 1878. The Leas then owned all of Roswell, as this version goes. Actually, the deed records show that is the date on which Wildy bought seven tracts of land totaling 440 acres from Marion Turner. The records do not show that Wildy ever transferred the land to Sallie, although she and her husband may have overseen much of it.

Some of the deeds refer to Major Wildy as a farmer and resident of Lincoln County. Together with other evidence, this suggests that he and some of his other children spent time in Roswell developing his lands. Upson, in the 1878 letter to his sister, states, "... We have a colony of Mississippi and Missouri people here at Roswell now, wealthy, intelligent, and educated ... . Several from here have gone back to Mississippi to harvest their cotton. They come from Satartia, Yazoo County and Warrenton." Just who these "several" were is not stated, but they may have included Major Wildy; his two sons, John, 24, and Ernest, 22; his daughter, Ada Holloman, 27, who was a widow; and his son-in-law, James Bonney, who was married to Mattie Wildy. Another source (Joseph D. Lea, *Roswell Register-Tribune,* October 11, 1906) confirms that two of those who returned to Mississippi to harvest cotton were Ernest Wildy and C.D. Bonney (James' younger brother), both of whom would return later.

Regional Leader

Lea was rapidly elevated to the status of community and regional seader soon after his arrival in the county. In the fall elections of 1878, he offered himself as a Democratic candidate for the Lincoln County Commission. Despite his newcomer status, he was elected.

Early Irrigation. While the Lincoln County War was in progress, as well as the years following the war in which Billy the Kid was such a thorn in the flesh of New Mexicans, Lea tried his best to continue his development of Roswell and the Pecos Valley. One important area was that of irrigation ditches or canals.

When Hiram C. Fellows conducted the first surveys of the present site of Roswell in 1866-67, he found an *acequia* coming from the North Spring River and crossing the area where the Chaves County Courthouse would be built later. Fellows also reported that a significant amount of land by the Hondo had been cultivated. No one knows who built that ditch, or when; but it is assumed that it was done by early Hispanic settlers. Van C. Smith may well have made use of the original ditch and even expanded upon it as he brought water from the North Spring River by *acequias* in front of and behind his buildings. Shinkle makes no mention of Smith's ditches in his book on Roswell's history.

Upson wrote to his relatives in the East on August 30, 1876, and describes the water resources of the Roswell area:

> The Hondo is south of the houses: northwest of the houses is the North Spring River about 1,000 yards distant. This river is as transparent as crystal and about 40 feet wide. The house is only two miles from the rise in the river and it is only 4 miles from the house to the mouth—it empties into ... the Hondo about 1½ miles from the house, and the Hondo empties into the Pecos about 2½ from the mouth of North Spring River, making North Spring some 2½ miles away. Besides North Spring River there is South Spring River which has its rise just four miles south of this house, and makes its juncture with the Hondo at its mouth, where they both, or rather all three empty into the Pecos. Besides these four rivers there are two smaller ones, their rise being from springs not more than 2½ and 3½ miles below the mouth of the Hondo. Six rivers within four miles of our door—literally alive, all of them with fish. Catfish, sunfish, bull pouts, suckers, eels, and in the two Spring Rivers and the two Berrendos splendid bass. These four rivers are so pellucid that you can discern the smallest object at their greatest depth. The Hondo is opaque, and the Pecos is so red with mud that any object is obscured as soon as it strikes water….

Because of the abundant supply, farmers built ditches to appropriate the water from the various streams to irrigate their crops after the beginning of Anglo settlement in the 1870s. According to Morgan Nelson, water history expert, some 160

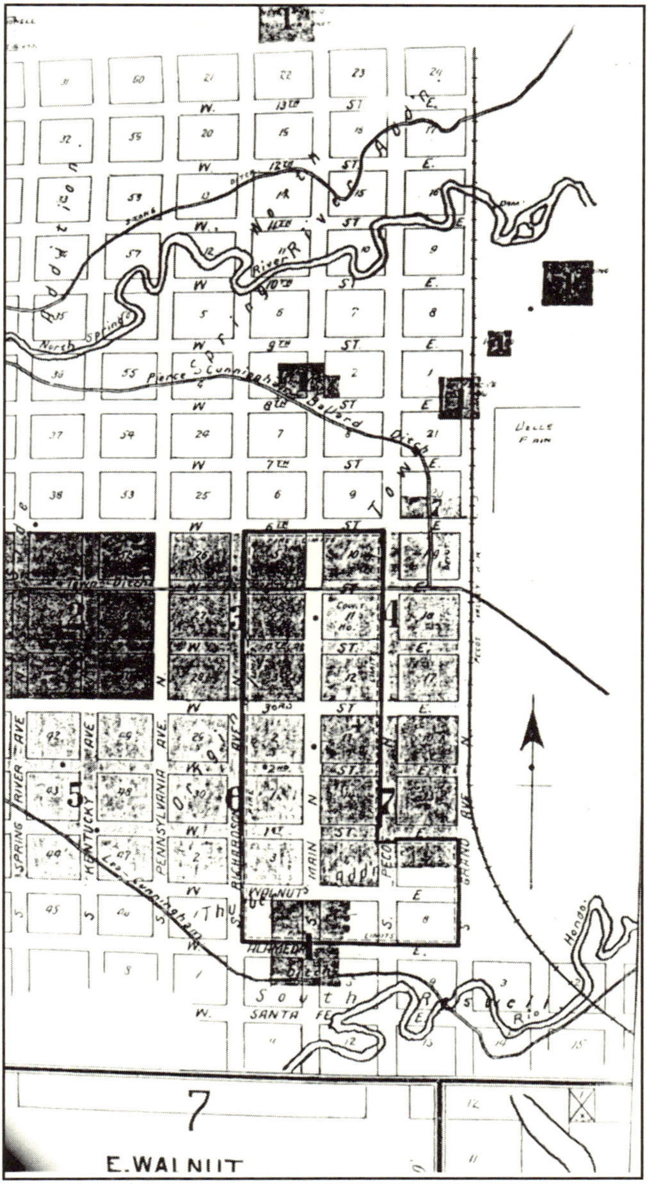

1905 Sanford Fire Insurance map of downtown Roswell, showing streams and irrigation ditches. Top to bottom: Stone Ditch, North Spring River, Pierce-Cunningham Ditch, Town Ditch, Lea-Cunningham Ditch, Río Hondo.
Courtesy: HCSNM

ditches have been identified in the Roswell area. Some of the first ditches were from the South Spring River in the area around the headquarters of John Chisum's South Spring River Ranch. The earliest legal claim to South Spring water was in 1872 for what came to be called the Woodlawn Ditch, followed by the Chisum Ditch (or Jinglebob Ditch) in 1876. These ditches watered new farmland a few miles southeast of Roswell.

J.C. Lea took preemptory steps to ensure that his lands would have access to water as they were developed. Apparently, in 1879, Lea began to build diversion dams to appropriate water from the North Spring River, which flowed just north of the townsite. In the summer of 1880, he filed papers with the Secretary of the Territory regarding three ditches. A dam was built at the "falls" of the North Spring, "... immediately above the main road leading from Roswell, New Mexico, to Las Vegas, New Mexico," according to a document dated September 1, 1880. Two ditches were to be taken at that point, both on the north side of the river. The larger ditch (ten feet wide and two feet deep) would follow an easterly course for about three miles, where it would cross the Hondo in a flume and intersect with another ditch. The water was to be used "... for agricultural purposes and for manufacturing purposes, to be used as the motive power for a Flouring and Grist Mill." The other ditch was to start near the same place but below the falls; it was smaller: six feet wide and two feet deep. Records and maps indicate that not all of the proposed venture was built, but it does seem likely that part of it was built about 1882-83 and was called the 'Roswell Mill Ditch.' It was also called the 'Blashek Ditch' because it was used to operate George Blashek's gristmill northeast of the townsite. The diversion point for the mill ditch was just east of the present railroad right-of-way.

Lea built a dam on the river a mile or so west of present Main Street about 1879 and took a ditch from the south bank of the North Spring River. He filed papers for the ditch in August and September 1880. The ditch tracked in a southeasterly direction for about four miles, crossing present South Main Street just south of present Alameda Avenue and crossing the Río Hondo in a flume in the 200-block of present East Tilden Avenue. The ditch was first described as ten feet wide and two feet deep, but the second document stated that it was sixteen feet wide. This ditch was called the North Spring River Ditch or the Lea-Cunningham Ditch because James and Harriet Cunningham had partial ownership. Many others eventually owned shares of the water. It was actually built about a half-mile south of the course described in the original documents.

Another important ditch that ran through the Roswell townsite was the North Spring River Centre [sic] Ditch. Lea and his sister Ella Calfee, together with the famous former Lincoln County Sheriff Pat F. Garrett and his friend and former deputy Barney Mason, incorporated this company on October 1, 1881. The point of diversion for Centre Ditch was described as "... 50 yds. above the road leading from Roswell, N. M. to the head waters of the river Berendo" [North Main Street]. It seems the dam was actually built farther west than described. According to other documents, it was diverted at the bend just north of present West Eighth Street and west of present North Missouri Avenue. Unfortunately, Lucius Dills, in an unpublished manuscript in the archives of the Historical Center for Southeast New Mexico in Roswell, states that the Eighth-and-Missouri location is the point of diversion for the Lea-Cunningham Ditch instead of the Centre Ditch; and that incorrect information has been used in some water studies. The course of the Centre Ditch was generally eastward for about four miles. It crossed present North Main Street at Eighth Street. The earliest plats of Roswell (1885) show the Centre Ditch and the Lea-Cunningham Ditch, as well as the original ditch across the present courthouse square.

Mescalero Depredations

The Mescalero Apaches carried out depredations in Southeast New Mexico during the years following the climax of the Lincoln County War. Around June 1879, they began raiding the Pecos Valley regularly about once a week. There was a let-up in August and September 1880, but then the Indians raided the Seven

Rivers area again. Local settlers looked to Captain Lea for help. In a letter to Governor W.G. Ritch on October 8, 1880, Lea reported, "... we were visited by two marauding bands driving off both cattle & horses, killing many of both, plundering houses & attempting the lives of our best citizens." The soldiers of Fort Stanton were no help in the Pecos Valley because they were kept too far west. In his appeal to the governor, Lea wrote:

> Now what we want is to have soldiers stationed permanently on the Pecos for our protection against the Indians. This country is filling up rapidly with new people. We have a thriving, prosperous settlement at Seven Rivers & Roswell & could we get protection from Indians a few months, I think we then could be strong enough to take care of ourselves.
>
> In addition to the U. S. Soldiers, we would like for you to authorize the organization, arming, and equiping of the militia in order that we might cooperate with the military in the protection of our lives & property .... It will require two companies of U. S. Soldiers Stationed at different points on or near the Pecos in this county ... . It is a large area & so thinly settled, which gives the Indians great advantage in their Devilish work … .
>
> Trusting that you will kindly consider the seeming presumption of my writing you on this subject without having the honor of your personal acquaintance & that this distressed people may hear favorably, & believing that you will do all in your power for us. I have the honor to Subscribe Myself
>
> Your Humble Obt. Sv't.
>
> J.C.Lea

Whether or not the governor did as Lea requested is not clear. It is unlikely that a militia was organized at that time, although the citizens no doubt continued to resist the Mescaleros on their own.

6

J.C. LEA AND THE LINCOLN COUNTY WAR

Upon settling in Roswell, Lea soon discovered that he had landed in a hornet's nest. There was a constant heightening of the tensions which led to the most noted event in the history of Southeast New Mexico: the Lincoln County War. The large county had originally been formed from the eastern half of Socorro County in 1869 and named after the recently-assassinated President Abraham Lincoln. In 1878, the area south to the Texas line, the eastern part of Doña Ana County, was added to Lincoln County and thus created the largest county in the nation, comprising some twenty to twenty-five percent of the area of New Mexico Territory.

Generally speaking, the Lincoln County War was a struggle between two factions who wanted to control the economy and the politics of the county. One camp is usually identified as the Lawrence G. Murphy-James J. Dolan party, which was affiliated with the so-called 'Santa Fe Ring' under the leadership of U.S. District Attorney Thomas B. Catron. The other was the John Henry Tunstall-Alexander A. McSween faction. The main commodities of value in Lincoln County were the government contracts to supply beef and other staples to the federal installations at Fort Stanton and the Mescalero Apache Indian Reservation.

Both factions were comprised of businessmen-ranchers, their workers, certain territorial and local office-holders, and assorted others. Both sides had the support of particular politicians, some Anglos, and some Hispanics. Some writers have called it a 'range war' or tried to categorize the two sides in various ways: newcomers against old-timers, Anglos against Hispanics, Irish against English, or Catholics against Protestants. None of these designations has much validity. A more accurate depiction would be: greed for wealth and power *versus* greed for wealth and power.

Each faction operated a big mercantile store in Lincoln. Each had a group of gunfighters who were mostly just local farmers and cowboys, but included some known criminal elements. The Murphy-Dolan gunfighters were more-or-less a sheriff's posse, and the Tunstall-McSween warriors were originally a constable's posse who had the power to serve arrest warrants that were issued by Justice-of-the-Peace John B. 'Squire' Wilson. At times, both posses had warrants for the arrest of several members of the other posse!

At the time of the Leas' arrival, others were departing Lincoln County because of the mounting violence. Lea was friendly with cattle baron John S. Chisum, who was a partisan of Tunstall and McSween. Lea had a lot of money tied up in his place, and he did not want to take sides lest he offend prospective settlers and customers, but, without law and order, the area would never be settled and developed.

Lea served notice to all parties that Roswell was neutral territory—no violence between the Murphy-Dolan and Tunstall-McSween factions would be allowed. According to Georgia B. Redfield in a manuscript dated January 27, 1938, Lea sent for William H. 'Kid' Bonney and laid down the law to him. "Bonney, if I ever catch you here in Roswell cutting up any of your capers, I'll take my Winchester and fill you full of holes."

The Kid replied, "All right, Cap'in. I promise I won't ever cut up any capers in your Roswell." And he kept his promise.

An article probably written by Postmaster Ash Upson for the April 29, 1892, issue of the *Roswell Record,* declares that Upson:

> ... warned them all that so soon as the first brawl occurred, the first gun fired, or the premises in any manner made the scene of battle, he would, at once, send his resignation to Washington, and his post office paraphernalia to Fort Stanton ... . Men who were fighting ... on both sides, depended upon the post office at Roswell for their intercourse with the world outside a very dangerous circle, and could not afford to risk its suspension. Hence,

> both factions respected the post office and the postmaster. The result was that never, during the years when Roswell was the center of an immense battlefield, rampant with anarchy, murder, arson, robbery and worse, not a shot was fired in malice or anger within four miles of Roswell ... .

The firm stands taken by Lea and Upson seem to have worked.

Roswell was touched, nevertheless, by the activities of some of the participants in the war. Following the killing of John Henry Tunstall on February 18, 1878, by Sheriff William Brady's posse, some of Tunstall's employees and friends formed themselves into a constable's posse to avenge his death. Later, the legal basis for their existence was removed so they became a sort of vigilante group, calling themselves the Regulators. They went to Dolan's cow camp on the lower Pecos where they arrested Frank Baker and William S. 'Buck' Morton, two members of the sub-posse that had killed Tunstall. The Regulators stopped by Chisum's South Spring River Ranch with their prisoners on the night of March 8, 1878, on their way to Lincoln. Morton, fearful that he and Baker would not survive the trip to Lincoln, wrote a letter to his cousin in Virginia asking him to alert federal authorities. The next morning, according to Robert M. Utley, in *Billy the Kid: A Short and Violent Life* (1989), "... the Regulators rode over to Roswell and reined in at Captain Lea's store, which also housed the post office. Morton entrusted his letter to Upson ... ." Sure enough, that night (or possibly the following night) at Agua Negra Spring northwest of Roswell, the Regulators found cause to shoot Morton and Baker for attempting to escape, as well as Robert McCloskey, who was supposedly along to see that the prisoners came to no harm.

On July 4, 1878, just two weeks before the climactic Five-Day Battle in Lincoln, the Regulators were enjoying the holiday at Chisum's ranch. Billy Bonney was "fascinated" by Sallie Chisum, John Chisum's teenage niece, "... and she returned the favor," according to Utley. "During the morning, accompanied by Frank and George Coe and several others, Billy rode over to Captain Lea's store to buy candy for Sallie from clerk Upson. On the return, twelve Seven Rivers possemen galloped to the attack ... ."

The sheriff's men and the Regulators fought a running gun battle from Roswell back to Chisum's ranch, and a standoff lasted all day until the deputies finally left. No one was hurt in the fracas.

Another incident during the war concerned Pitser Chisum, John's brother. According to a quotation in Sophie Poe's book, *Buckboard Day* (1936), John Chisum told her that some of the Murphy-Dolan faction planned to mutilate Pitser.

> They were going to capture him at Captain Lea's house, where he was visiting, and give his ears a jingle-bob slit and put a Chisum Long Rail brand on him. The 'posse' was all set for the performance. They got together back of Captain Lea's store, and tanked up on whisky; so they were all ready for anything.
>
> Captain Lea got wind of it through an outsider. He sent his colored boy out to the corral to get Pitser's horse, and take it secretly to the back door. There he helped my brother to get out and mount ... . Luckily, it was a dark night, so he got away without much trouble ... . The posse heard his horse's hoofs hitting the trail, and sent a few shots after him ... .

From July 14th to the 19th, McSween and his men fought the 'Five-Day Battle' in Lincoln against Dolan and the sheriff's men. It all ended when the deputies set fire to the McSween house; McSween and others died trying to escape the burning house.

Aftermath of the Lincoln County War

The Five-Day Battle did not bring peace to Lincoln County. The war gradually diminished, but disturbances in Lincoln County actually escalated over the next few months. With a lack of effective law enforcement, several new outlaw gangs arrived to terrorize the county.

Trouble on the Berrendo. The desperate outlaws particularly targeted Hispanic settlers. Despite his efforts for neutrality and peace, Captain Lea was caught in the middle of some of the violence. Lea's contemporary, James M. Miller, wrote an account of the event. Some of the vicious criminals forced many Hispanic settlers to abandon their homes, and many of them returned to the

Rio Grande Valley whence they came. About ten families moved in with a little settlement on the Río Berrendo north of Roswell. About twenty of the outlaws showed up and stole the livestock and everything else the Hispanics had that was worth carrying away, "... even the women's' shawls." The outlaws ordered the Hispanics to leave or be killed. The settlers were so outraged and indignant that some of them talked of killing all the Anglo settlers around Roswell.

A few days later, Miller writes, Captain Lea and Judge Edmund Stone were heading for their sheep camp some fifty miles north of Roswell. Arriving at the Berrendo, they saw the Hispanic families loaded in about a dozen wagons in preparation for abandoning the settlement. They used their wagons to block the road at the river so that Lea and Stone could not proceed. The Hispanics began to threaten Captain Lea, claiming he had some responsibility for their recent losses, and called him a "damned Texan." They did not rob Lea nor take his horses, but one of the 'Mexicans' jerked out his six-shooter. At that point, another of the settlers, Abrán Candelario, recognized the Captain and knew him to be friendly and peaceful. Candelario, who was a very large and strong man, got between the would-be shooter and Captain Lea. He grabbed the armed man in a bear hug "... just in time to save Lea's life," according to Miller. Sophie Poe wrote of the event that Candelario yelled at the rifleman, "This is my *buen amigo.* You can't kill him until you have killed me." Lea and Stone were able to continue on their way unharmed, thanks to the intervention and persuasion of Candelario.

Commissioner Lea. Many settlers—Hispanic and Anglo—left the county, some never to return. John Chisum and his brothers moved their herd temporarily to the Texas Panhandle. Lea's election to the Lincoln County Commission in the fall of 1878 put him in a position of influence and power to affect efforts to clean up the county.

James M. Miller, in an article he wrote for the March 17, 1918, *Roswell Evening News,* states:

> When Captain Lea took charge, ... a favorable change took place .... The place was transformed from a rendezvous for the

> Murphy-Dolan outlaws to a safe and happy home, not only for the Lea family but for their friends; and for many years there were no murders committed in Roswell ... .

Lea was not content to be neutral and did all he could to bring an end to the violence.

Buffalo Soldiers in Roswell. When the federal government changed the rules and allowed the military to help civil officers keep the peace, Captain Lea took advantage of the new situation and had Upson write to the commander of Fort Stanton, Colonel Nathan Dudley, and ask him to send troops to protect the post office at Roswell and the mail carriers. Lea himself wrote a similar letter because of the incident on the Berrendo. Dudley promptly dispatched Company 'F' of the Ninth U.S. Cavalry commanded by Captain Henry Carroll. The black 'Buffalo Soldiers,' the first military ever to be stationed in Roswell, were encamped in the town from early November 1879 until early February 1880. According to Ernestine Chesser Williams in *Treasures of History III* (1995), "Some old-timers say the camp was located northwest of downtown Roswell on the banks of North Spring River." Mrs. Williams also states that some of the Southern women in Roswell resented being protected by Union soldiers, especially black troopers. There is no record that the Buffalo Soldiers engaged in any inappropriate conduct.

Captain Carroll had little to do, M.G. Fulton avers in *Roswell In Its Early Years* (1963), other than patrolling the roads to protect the mail. He did report what he claimed was the "... first attempt at enforcing civil law ever known in that precinct." Two drunks got in a fight in front of Captain Lea's store. One named Waldo knifed the other, whose name was William Smith. The local constable hauled him before the justice-of-the-peace, who fined Waldo $2.50, even though this was not an appropriate case for his court. Carroll wrote, "A charge with the intent-to-kill was not a case for a justice of peace. But a little whiskey supplied any authority that might have been wanting in the case." Neither Carroll's report nor the Lincoln County records reveal the names of the constable and justice-of-the-peace, but they were likely A.J.

Lawton and A.H. Whetstone, respectively. These two were elected or re-elected in 1880.

Later, Captain Carroll's men retrieved Susan McSween's cattle, which rustlers had driven off their Río Feliz range and concealed in Rocky Arroyo (near present Carlsbad). They brought the herd to Roswell, where Mrs. McSween's representative took charge and drove them back to Lincoln.

From an economic standpoint, the sojourn of the Buffalo Soldiers was beneficial to Roswell. The military trade was a bonanza for Lea's store, and the farmers in the area also prospered. The amount of supplies needed for the soldiers and their mounts was considerable.

Nevertheless, not everyone was happy with the military occupation of Roswell. In the 1892 article, Upson (or whoever wrote it) states,

> It is not sure that they (the soldiers) were of any benefit to honest people. There was no fault to be found [in] the colored soldiers, but Capt. Carroll resented being called a soldier, because he was an officer. This officer, "who was no soldier," appeared to be spending his energy in the service of anyone except the government and the army—a partisan, he labored for any party, or faction, which his prejudice or his interest influenced him to favor and embrace. It is very probable that the strenuous exertions of Carroll's company tended to lengthen the struggle between better men than he. There is little doubt that Capt. Lea and his law-abiding friends felt that in avoiding the lawless Scylla, they had fallen into the legalized Charybdis. The soldiers were withdrawn, much to the relief of the Valley ... .

Getting Rid of the Kid. Although Captain Lea was more-or-less neutral during the Lincoln County War itself, he was deeply involved in its aftermath. He especially had a hand in taking steps to rid the county of a friendly, polite, boyish, yet ruthless killer and rustler, who at the time called himself 'William H. Bonney.' By the end of 1880, he was being referred to as 'Billy the Kid' and today is known throughout the world by that name.

The Kid had surrendered voluntarily to house arrest at Lincoln for a few months in the spring of 1879 while he was testifying in court. When he realized in June that the pardon Governor Lew. Wallace had promised him was not going to happen, he "went to the mountains" and began a year and a half of rustling activity along with a few of the former Regulators and some other rowdies who had now entered the cattle-rustling business.

Even though the settlement of Roswell itself was not the scene of overt outlaw activity, the gangs of outlaws and rustlers controlled the county outside of Roswell. The outlaws were not only busy stealing large herds of cattle, but they also had the effrontery to rob the U.S. mail and to hold up the stagecoaches coming out of Las Vegas. According to Padgitt, Captain Lea was "... held up in his buckboard and had his Winchester taken from him."

Lea realized that his vision for Roswell as the center of a prosperous area, served by a railroad, and encouraged by himself and other entrepreneurs, was never going to materialize unless something could be done about the lawlessness rampant in the country. In the summer of 1879, several leading men of Lincoln County became convinced that the only way to solve the problem was to eliminate the Kid. Captain Lea, J.J. Dolan, Judge Warren Bristol, and finally John S. Chisum, as well as other men in the Territory, concluded they must find someone who was willing to go after the Kid and kill him, according to Jon Tuska in *Billy the Kid: His Life and Legend* (1994). They needed someone from outside of Lincoln County who had not been a partisan in the Lincoln County War.

Patrick F. Garrett, a former Texas buffalo hunter who had settled in Fort Sumner, was the man whom Lea and Chisum chose. Garrett cultivated his acquaintance with Captain Lea, John Chisum, and other leading cattlemen in the area. George Kimbrell, Republican, was up for reelection as sheriff of Lincoln County in 1880. He thought he had the position secure and expected to be nominated by both political parties. However, Kimbrell showed no inclination to even pursue, let alone kill the outlaws. Garrett satisfied Lea and Chisum that he would, if elected sheriff, hunt down the Kid and any other cattle rustler and dispatch them to their Final Reward. Captain Lea helped Garrett take up some good farmland east of Roswell so he could qualify for nomination by

Patrick F. Garrett, Sheriff of Lincoln County and Deputy U.S. Marshal.
Courtesy: Rose Collection #1781.
Western History Collections,
University of Oklahoma Library

the Democrats as a resident of Lincoln County. Lea, Chisum, Dolan, and other men of substance supported Garrett's nomination because of his promise to kill outlaws. "It seemed for a brief, fluttering moment," Tuska writes, "as if Lincoln County was at last on the verge of a majestic consensus of the people—those with money and influence, anyhow." Nolan asserts (e-mail letter, June 7, 2001) that the men were "being entirely pragmatic" in choosing Garrett for sheriff; there is scant evidence to indicate that they were fond of Garrett.

Captain Lea was in a good position to influence Garrett's election because he represented the Pecos Valley and Roswell area on the Lincoln County Commission. He could produce the votes of his area because of his prominence and influence, although he did not stand for re-election as a Lincoln County Commissioner in 1880. Further, he and Chisum had sway over the big cattlemen, while Catron and Dolan together were able to deliver the votes of many men who owed them money or favors. With the support of these powerful opinion-leaders, Garrett won the election on November 2, 1880, with 320 votes to Kimbrell's 179.

It would be two months before Garrett could take office on January 1, but Captain Lea and the others did not want him to wait that long. Accordingly, Kimbrell appointed him a deputy; and Garrett, for all practical purposes, became the acting sheriff. He needed someone to handle his paperwork, so Upson in Roswell was appointed as a 'clerical' deputy. In addition, a federal agent presently made Garrett a deputy U.S. marshal in order for him to be able to act outside of Lincoln County if necessary.

According to Frederick Nolan in *The West of Billy the Kid* (1998), that federal agent was Azariah F. Wild of New Orleans, who was a special operative of the U.S. Secret Service. He had come to the county because Captain Lea, J.J. Dolan, Ira Leonard, and others had written to the U.S. Department of the Treasury to inform them that someone was passing counterfeit one-hundred-dollar bills in White Oaks and Lincoln. Wild assumed, probably incorrectly, that Billy the Kid was behind it all. There is no question that certain of the rustlers were involved in passing the bills, but some of them may have been duped as well.

Lea and Chisum saw Wild's assignment as a possible opportunity to get him to help Garrett rid Lincoln County of the rustlers, according to Robert M. Utley in *Billy the Kid: A Short and Violent Life* (1989). Together with Lea and Garrett, Wild started immediately after Garrett's election to devise a daring plan to raid the outlaw hangouts around Fort Sumner. Lea and Garrett agreed to organize a posse of fifty men, which Garrett and Deputy Robert Olinger would lead. At Fort Sumner, Frank Stewart and a group of men from the Texas Panhandle who had been hired to come to New Mexico Territory to recover stolen cattle would help the posse with the sweep.

As a precursor to the posse's raid, Wild wrote to Lea, he needed a trustworthy, cautious man to snoop around Fort Sumner and find out where the counterfeiters were likely to be hiding. Lea and Garrett both vouched for the loyalty and honesty of Barney Mason of Fort Sumner, so Wild hired him. Tuska writes that Garrett and Mason probably did not accept Wild's belief that Billy the Kid was involved in the counterfeiting operation, but "... they were willing to use (Wild's) money and influence to finance their operations ..." against the rustlers. Wild appointed Olinger a deputy U.S. marshal, which enabled him to arrest criminals in Fort Sumner, San Miguel County. According to Tuska, Wild even had Captain Lea appointed a deputy U.S. marshal about January 2, 1881.

A heavy snowstorm on November 14 paralyzed the country, according to Nolan, requiring a delay in the plan. Wild and fifteen men were about to depart White Oaks on November 19 when he received a message from Captain Lea that Garrett was on the way there with Mason. According to Garrett in his book, *The Authentic Life of Billy, the Kid* (1882), Garrett stopped in Lincoln and Mason went on to White Oaks to try to infiltrate the outlaw elements there.

After a few days, Garrett and Mason both returned to Garrett's farm east of Roswell. Garrett writes that he heard daily reports of criminal activity by 'The Kid' around White Oaks. He heard that the gang had moved to Bosque Grande north of Roswell, so he summoned his neighbors to form a posse. On the way to town,

Garrett and Mason spotted a fugitive heading toward Chisum's South Spring River Ranch. They chased him and arrested him. He was a man named 'Cook'; Nolan states that the man may have been either Sam Cook, who had been charged with cattle rustling at Las Vegas in August 1879, or Joseph Cook, who was a cattle rustler from Texas whom Garrett later apprehended. At any rate, Garrett took Cook to Roswell, put him in irons, and turned him over to Captain Lea. Garrett states that Lea was in charge of Cook for some three to four weeks.

The well-known story of how Garrett and his posse caught and captured Billy the Kid and his men need not be detailed here, but some aspects of the effort affected Roswell. There was some panic in both White Oaks and Roswell while the Kid's gang was on the loose. A lot of excitement was stirred up in White Oaks when Deputy Sheriff James Redman claimed the Kid had come into town and tried to assassinate him. At the same time, Wild was in Roswell holding Joe Cook. Folks around Roswell heard that the Kid's gang was nearby and would attempt to extricate Cook from the hands of the law. The pitch of the rumors became higher and higher as people gathered around Captain Lea's store, getting ready for a final confrontation with the outlaws; but when nothing happened they quietly left. Garrett writes that Lea sent Cook to the jail at Lincoln, and Cook escaped from there.

Interestingly, while Captain J.C. Lea was instrumental in securing Garrett for the job of ridding the country of the Kid, in the final days Lea was also called upon to use his good offices on behalf of one of the outlaws. Charlie Bowdre presumably wanted to mend his ways and leave the 'owl hoot' trail, but he found it impossible to separate himself from the Kid and the gang.

Wild and Garrett's grand-sweep plan failed to materialize, and Garrett had to use a posse of a dozen or so men, his 'neighbors.' One of the possemen was Sallie Lea's young brother, Ernest L. Wildy, Nolan states. Garrett's account indicates that the posse arrived at Fort Sumner at daylight on November 27 (more-or-less; sources disagree on the dates of these events). There, Garrett "... received a letter from Capt. Lea, detailing further depredations of

The Kid and band about White Oaks, the killing of Carlyle, etc." Lea's reference was to the accidental shooting death of Jim Carlyle, leader of a posse out of White Oaks that was pursuing the Kid and his cohorts. At the Greathouse Ranch, Carlyle went inside to parley with the desperadoes. When he came crashing out a window, the posse thought he was the Kid making a break; so they shot him. After receiving the message from Captain Lea, Garrett and his posse immediately set out to scour the area around Fort Sumner for the outlaws.

On December 5, 1880, Garrett and the posse raided the Yerby Ranch in search of the desperados, but they found only Manuela Bowdre (Charlie's wife), another woman, two mules belonging to Bowdre and four stolen horses—which they confiscated. Some twelve miles east of Fort Sumner, Garrett and the posse stopped at the Wilcox-Brazil Ranch to eat. Erastus Wilcox gave Garrett a message that Bowdre wanted to negotiate with him. Garrett met with the outlaw about two miles out of Fort Sumner and showed him a written statement from Captain Lea to the effect that any of the Kid's gang who defected would receive consideration for leniency. Bowdre was between a rock and a hard place. He had been indicted for the April 1878 killing of Andrew 'Buckshot' Roberts at Blazer's Mill and therefore could not accept Governor Lew. Wallace's offer of amnesty to any outlaws who would stop the violence and tell what they knew in a court of law. He dared not turn against his fellow rustlers, lest they retaliate.

Bowdre decided to write an appeal to the good captain. His handwriting, spelling, and grammar suggest that he was considerably more intelligent and educated than most of his colleagues:

Fort Sumner Dec. 15th, 1880

Capt. Lea,

I have broke up housekeeping and am camping around, first one place & then another on the range, so that no one can say that Yerbys ranch is the stoping place for any one. So no party will have any excuse for going there unless they are after me. I thought this a duty due Mr. Yerby, for if there is nothing to eat at the ranch no one will go there & there will be no chance for a fight coming off there & Mr. Yerby's property injured. If I don't get clear I intend to leave some time this winter, for I don't intend to take any hand fighting the territory, for it is different thing from what the Lincoln Co war was. The only difference in my case & some of the officers in Lincoln Co. is that I had the misfortune to be indicted before the fight was over & did not come under the Gov's pardon. It seems to me that this would come to their minds once & awhile, when they are running around after me, but I suppose it is human nature to give a man a kick when you have the upper hand.

I saw the two Billies the other day & they say they are going to leave this country. That was my advice to them for I believe it is the best thing they can do.

Don't you think if you can get the Gov interested in my case, that it could be thrown out by the Dist Att'y without my appearing at court. I don't doubt your good intentions, but during the present state of the country I think there is some danger of mobing. For my name has been used in connection with a good many things that I have had nothing to do with, which outside parties can't know. I have taken your advice in regard to writing to Mr. Yerby & have no doubt but what he will do all he can for me. I know of nothing to urge in my favor, now than that others were pardoned for like offences. Experience is a good but slow teacher & I think if I keep my mind, I will let every man do his own fighting so far as I am concerned & I will do my own.

Respt
Chas. Bowdre.

Although Captain Lea had hardly met Bowdre, he was convinced that the outlaw's desire to go straight was sincere. On Christmas Eve, Lea took pen in hand in Roswell and wrote a four-page letter to the governor asking him to pardon Bowdre:

Roswell N.M.
Dec 24th 1880
To His Excellency Lew Wallace
Governor of N. M.

Dear Sir
I herewith enclose you letter from Charles Bowdre

John S. Chisum, 'Cattle King of the Pecos.'
Courtesy: Rose Collection #1601. Western History Collections, University of Oklahoma Library

Charlie Bowdre and his wife.
Courtesy: Haley Memorial Library and History Center, Midland, Texas

who was engaged in The Lincoln Co War, & for whom there has been a warrant for I suppose two years. He was I think indicted, previous to the date of your proclamation pardoning those who participated in these troubles. Since, he has lived in San Miguel Co near Fort Sumner.

Circumstances perhaps over which he had no control may have connected his name with lawless men. But from all the information that I have been able to get of Some of the best people of San Miguel Co he has been engaged at work in the employ of Mr. Yerby & at this time has full charge of Mr Yerby's stock range. I could not believe that Mr. Yerby would employ a man in that capacity who he did not believe was correct. I have been trying for several days to ascertain the character & Status of Bowdre in this Co & they are universally of the opinion (so far as I have been able to ascertain from the best & leading people of the Co) that he ought to be assisted in getting clear of this indictment. Bowdre has been writing to me since Garrett Started out to make the arrests of Bonney & he's (?) asking me to assist him &c & expresses a desire to get clear & to give himself up to the authorities & to make a good citizen & we believe that he will.

We further believe that if cleared of this, that he would be of invaluable service in clearing the country of bad characters. He is a man far above ordinary in point of intelligence. I have never seen him but twice in my life. But all who know him say you can depend upon his word & I have no doubt of it.

Taking into consideration all the circumstances, & the many difficulties that surround us, we think it decidedly best for the country & all concerned to try & clear him, & we hope if you agree with us that you will at once see the Dist'. Atty'. & have him promise to throw the case out of court. I am <u>sure</u> that every good man or every unprejudiced man who has the interest & welfare of our county at heart will concur in your actions.

His position now is that of a man between two fires. Garrett is after him with a warrant & he is afraid of the outlaws who infest that vicinity. We had better try to make a good man of him & we have many reasons for believing that we can.

Should we drive him to the outlaws, He would or could give them more strength than any other one man. They are now too much for our strength & resources. They have the best horses arms &c that money will buy & the finest field glasses in the country so that they & their spies always know when they are being hunted, & their friends in all parts of the country to post & assist them. –

Hoping to hear from you favorably in the case of Bowdre.

I Am Your Ob't. S'v't
J.C.Lea

Captain Lea enclosed Bowdre's letter with his own and mailed it to the governor. Both letters are on microfilm in the New Mexico State Records Center and Archives. But it was a forlorn hope. Lea was unaware of the activities of Pat Garrett and his small posse. The posse caught up with the Kid's gang at the old Fort Sumner military post, probably on December 19, although the chronology of these events is hopelessly muddled, according to Utley. Tom O'Foliard was killed in that encounter, but the rest of the gang escaped. Garrett and his men found them again at Stinking Springs, east of Fort Sumner, at 3:00 a.m. on the (likely) 23rd. At dawn, Bowdre emerged; and Garrett, thinking Bowdre was the Kid, ordered his men to fire. Bowdre was fatally wounded, but according to most accounts he came out to Garrett's position before he died. The posse arrested the Kid and the others and carried Bowdre's body to his broken hearted wife.

So, at the exact moment that Captain Lea sat at his desk penning a letter on behalf of Bowdre on that Christmas Eve, "... in the abandoned hospital building at Fort Sumner, the grieving Manuela Bowdre dressed her dead husband in a new suit paid for by Pat Garrett, laid out his body, and encircled it with glowing candles," Utley writes. At the same time, Pat Garrett put Billy the Kid and his other prisoners in a wagon and started out from Fort Sumner on a long, cold journey to the Las Vegas jail.

Billy the Kid was moved to the jail at Santa Fe for a while, then transferred on a change of venue to La Mesilla for trial. He was found guilty of killing Sheriff William Brady and was to be hanged in Lincoln on May 13, 1881. He cheated the rope, however, and escaped from the Lincoln County Courthouse for another few months of freedom. Garrett again had to hunt him down.

* * *

On June 4, 1881, the Bonney brothers—James, John, and C. D., Jr.; no relation to William H. Bonney—arrived in Roswell from Yazoo County, Mississippi. Young C.D. promptly bought into Captain Lea's store. Less than six weeks after he started working in the Lea, Bonney & Company store, Bonney sold ammunition to Pat Garrett and his deputies John W. Poe and Thomas C. 'Kip' McKinney as they started on the junket which ended in the shooting of William H. 'Billy the Kid' Bonney. Garrett again found the Kid at Fort Sumner and shot him to death on July 14, 1881.

In the year following the death of Billy the Kid, Upson ghostwrote Pat Garrett's famous little book, *The Authentic Life of Billy, the Kid.* According to Frederick Nolan, in his annotated edition, Garrett's book set the 'facts' of Billy the Kid, although hardly any of it stands up to the scrutiny of modern research. Upson fabricated a background for the Kid, while glossing over anything in Garrett's dealings with the young outlaw that might have made Garrett look bad.

Some writers have confused Captain J.C. Lea with others who had a part in Lincoln County's history. His brother, Frank H. Lea, was a justice-of-the-peace in White Oaks in the early 1880s during the campaign to catch Billy the Kid and other outlaws. Shortly after the Kid's escape, Jesse Smith Lea, a cousin of J.C., arrived in Roswell on May 3, 1881. In July, Lea became a deputy and the jailer at Lincoln, where he served five years under both Garrett and Poe. In 1885, he killed a prisoner who was trying to escape. Smith Lea later became a businessman in Roswell and held many city and county offices.

7

DEVELOPMENT OF ROSWELL TOWN IN THE 1880S

Major William W. Wildy died of 'dropsy of the heart' at Churchill, the plantation home of his daughter and her husband, Mattie and James Bonney, in Yazoo County, Mississippi, on February 7, 1881. When his wife Laurena died in 1872, she was buried in the cemetery at the little Methodist church in Phoenix, according to Barbara Buckner of Carlsbad, great-granddaughter of the Wildys. The family assumes that Major Wildy is buried next to his wife, but there is no record or marker to that effect. Lawrence Long of Yazoo City believes Wildy may well be buried in an unmarked grave at Churchill Plantation, which Long presently owns. Long is a descendant of Laurena Wildy and her first husband, Judge Elijah Robertson Childress.

When W. W. Wildy's estate was settled, Sallie Lea secured the title to two hundred acres of her father's property, including that on which the Roswell store and hotel were located, on November 20, 1883, when she paid $2,400 to her brothers and sisters for their interest in it. The siblings retained the other tracts.

Captain J.C. Lea had been acquiring land since his arrival in the Pecos Valley. Between 1879 and 1885, he secured patents on over 1,200 acres within the future townsite of Roswell. These acquisitions placed him in an opportune position for developing his town.

Mrs. Buckner relates a family story. In the early days of Captain Lea's cattle operations, Ernest and John Wildy worked for him as cowboys. They were out west of Roswell when one of the other cowboys died. They buried him on top of a small hill and returned to town to report the death and to get a replacement.

When they came back, as they neared the little hill they thought their eyes were playing tricks on them. It appeared that someone was on the cowboy's grave waving his arm at them. When they got to the grave they saw that a coyote was trying to dig up the body, and what they thought was a man's arm waving was actually the coyote's tail.

Judge Frank H. Lea

Some of Captain Lea's relatives also made their way to Roswell at an early date. The first was his younger brother, Frank Houston Lea. After the Civil War, Frank, as well as J.C., had spent time in Louisiana. There he and Sue Whetstone were married at 'Auburn' —her parents' plantation near Bastrop in Moorehouse Parish on November 14, 1866. They lived at Auburn for a few years and had two children there: Joseph D. and Minnie.

In 1871, Frank and Sue moved their family to his hometown, Lee's Summit, Missouri. Other children were born to them in Missouri: Carrie, Jennie, and Gertrude. Since Frank's brother J.C. and Sue's brother Asbury Whetstone both lived in Roswell, Frank and Sue decided to move to the frontier in 1879. They put everything on a train to Las Vegas. Whetstone met them with an ox-drawn wagon and a hack. Frank Lea's daughter, Gertrude, wrote in 1937:

> The journey across the plains was hard and tiresome—nothing but miles and miles of barren prairie with no houses only those of Fort Sumner between Las Vegas and Roswell to break the monotony. When we drew near to Roswell and crossed North Spring River we thought it a beautiful stream and that we had reached the 'Promised Land.'

There was nothing to Roswell at that time except the boarding house/residence of Captain Lea and his store/post office, both located where Pioneer Plaza now stands (400-block of North Main Street). Gertrude stated, "Our entire family occupied one room in the store, while a house and a store were being built for us in White Oaks."

At Frank H. Lea's home, 1897. Joe Lea holding Gertrude Joyner, Frank H. Lea, Minnie Lea Minter, and Mary Lea Joyner. *Courtesy: HCSNM, Photo No. 462-D*

Judge Frank H. Lea family and relatives, July 1888. This house was located between Main Street and Richardson Avenue (facing east), and between Second and Third Street. Standing, *l. to r.*: Gertrude Lea, Sam Parsons (Baton Rouge, La.), Mary Lea, Asbury Whetstone (Bastrop, La.). Sitting, *l. to r.*: Pearl Lea, Park Lea, Doss Whetstone (Oakridge, La.), Joseph D. Lea, Ralph Parsons, Mack Minter, Jennie Lea. Top row sitting, *l. to r.*: Carrie Lea, Frank H. Lea, Susan Caroline Whetstone Lea. *Courtesy: HCSNM, Photo No. 462-C*

According to the *Roswell Daily Record* of February 10, 1905, Frank Lea was the "original settler" of White Oaks. He used a tape measure to stake off the town. Lea served as a Justice-of-the-Peace in White Oaks—the beginning of some thirty-five years of service in state and local government. Another child, Pearl, was born in White Oaks.

In 1884, Lea moved his family back to Roswell where he built one of the first houses in the town. After the streets and alleys were platted, the house fronted on the alley between Main Street and Richardson Avenue and north of West Second Street, where Lea also had a productive peach orchard. The adobe house washed down in the flood of 1904.

Ella Lea Calfee

Ella Lea Calfee, Captain Lea's younger sister, arrived in Roswell in 1880. She had married Dr. James H. Calfee in Tennessee, and their first son Frank L. Calfee was born there in 1871. Their other son, Edgar Calfee, was born in Arkansas about 1878. They then moved to Las Vegas, New Mexico Territory, where Dr. Calfee died in 1879. After Mrs. Calfee lost her husband, she and her two sons moved to Roswell in 1880 to be with the Leas.

Lea, Bonney & Company

C.D. Bonney, Jr., and his two brothers, John H. and James, all of Yazoo County, Mississippi, arrived in Roswell in a buckboard from the railroad at Las Vegas on June 4, 1881. The Bonney brothers were the sons of Dr. Caleb D. Bonney, Sr., a wealthy doctor who had plantations in Yazoo County, Mississippi, and Shelby County, Kentucky. James Bonney was Sallie Lea's brother-in-law, married in 1866 to her sister Mattie Wildy; he soon returned to Mississippi. John Bonney went to the Organ Mountains to prospect for gold and lived for a time in the mining boomtown of Organ and later in Las Cruces.

Twenty-one-year-old C.D. immediately bought an interest in Captain J.C. Lea's store, which then became "Lea, Bonney & Company." Their letterhead read, "LEA, BONNEY & Co., dealers in Staple and Fancy Groceries, Flour, Feed, Boots, Shoes, Hats, Caps, Gloves, Books and Stationery." Lea advised young Bonney, "Every man who comes to Lincoln County should make a decision early, either to carry a gun or never carry a gun." Lea never carried a six-shooter, so Bonney decided he would not carry one. He observed later that men who carried guns inevitably had to draw them. "I have seen several men dead as door nails, only because they drew too late," Bonney told his son, Cecil.

C.D. Bonney was an avid hunter and fisherman who loved to spend his free time in the outdoors of Southeast New Mexico. He related a story to Georgia Redfield in the 1930s about some of his early experiences. He said that in July 1882, he and others were sitting out in front of the Lea hotel when an orderly came in with a message from one Captain Scott, who was camped in an old corral near present Second and Main Streets. Scott had arrived from Fort Stanton with a cavalry outfit in search of a raiding band of Indians made up of both Mescaleros and Comanches. Bonney agreed to serve as a scout for Scott because he had seen the band about ten miles northeast of Roswell that morning. They left Roswell at 2:00 a.m. and caught up with the renegades at Mescalero Sands. In the shoot-out that ensued, Bonney stopped the leader of the Indians from killing Captain Scott. Bonney stated:

> We then disarmed all the Indians, who soon gave up, after we had caught their leader, and they had pretty good guns too. We went on over to Mescalero Springs and camped all night. We divided the Comanches from the Mescaleros and sent the Comanches to Fort Sill and took the Apaches back to Mescalero Reservation.

There are no known pictures or records to indicate whether Lea made any modifications to the store building that Smith & Wilburn had built. A story in the April 27, 1900, *Roswell Register* provides some information about the store:

> There were bullet marks in the double doors, where festive cowboys, 'tanked up' with 'red eye,' amused themselves in the

> early days making a target of them 'just for fun'.... While Captain Lea was in business there, George Griffin, who thought he was a bad man, threatened to kick the doors in one night, but changed his mind when he learned that Captain Lea and a double barreled shotgun were on the inside awaiting developments.

Sophie Alberding, 'Gal Tenderfoot'

Another early eyewitness account of Roswell is by Sophie Alberding Poe, who related her experiences in a book about her famous husband, John William Poe, in 1936: *Buckboard Days.* Her brother, Fred Alberding, had been a sheepherder for Captain Lea in Colfax County and had come to Roswell with the Leas and continued in the Captain's employ. Sophie was urged by their family to come from Illinois to visit him. Fred met her train at Las Vegas and brought her to Roswell in a buckboard. She is vague about most dates in her book, but it seems likely that she arrived in late 1881 or early 1882. She moved in with the Leas, to whom she was distantly related.

Mrs. Poe described Mrs. Lea as "... a woman of medium height, with a mass of lovely brown hair coiled on top of her head. She had the graciousness typical of the women of the South." She next met Ella Calfee. She writes:

> Mrs. Calfee was very much a contrast to [Mrs. Lea], in appearance. Like her brother, the Captain, she was tall and angular. Both women wore long, full calico dresses reaching to the ground, and were much beruffled around the lower edge. They did not make the least pretense of being stylishly dressed. They had been in New Mexico so long that they had lost interest in 'prevailing styles.' They wore what seemed best suited to their needs.

Mrs. Calfee and her sons occupied the upstairs room of the Lea residence, and the 'gal tenderfoot' Sophie Alberding moved in with them.

Mrs. Poe's description of her first meal at the Lea house provides clues to community life. The former Smith & Wilburn hotel, which had been a saloon and casino as well, was strictly a

Sophie Alberding Poe and her husband, Sheriff John W. Poe.
Courtesy: HCSNM, Photo No. 1206

boarding house under Mrs. Lea's supervision, with no liquor allowed. As long as Sallie Lea lived, no establishment selling liquor by the drink was permitted in Roswell. Some writers have claimed that she had the liquor ban written into her deeds, but the fact is she executed very few deeds; and no specific example of such a ban has been found. Mrs. Poe writes, quoting her brother Fred:

> Captain Lea, of course, keeps whisky in his store, for it's just as much a part of the supplies as sugar or flour or gunpowder. The cowboys wouldn't work if they couldn't look forward to a few drinks when they came to town. Captain Lea sells it by the bottle or by the jug, but he won't allow them to drink it in the store ... .

On her first morning in Roswell, Sophie Alberding came to the dining room to find twenty young men seated on benches on each side of a long, homemade table. She received many stares and much teasing from the men, who were "connected in one capacity or another with the Lea outfit." The number of diners at the Lea table suggests that Captain Lea ran a sizable livestock operation.

Because of the large number of travelers stopping in Roswell on their way to Las Vegas, Pecos City (Texas), White Oaks, Lincoln, or other destinations, Captain Lea felt compelled to provide lodging and meals for them in his home because there were no other such facilities. He also also charged a nominal fee for meals because of the large number of boarders frequenting the Lea residence.

Sophie Poe stated that across the street from the Lea house was a large corral with an adobe wall five feet high. Poe's description of the location of the corral seems to conflict with other accounts which place it behind the house. In one corner was a small house where her brother, Fred, lived. It was his job to protect Captain Lea's prize horses. In an article in *The Cattleman* (October 1946), Georgia Redfield stated, "The pride of Captain Lea's many possessions was Charley Wilks, a famous big bay, fine blooded stallion—that sired a better stock of horses in the Pecos Valley of New Mexico."

J.C. and Sallie Lea's second child was born in the boarding house/residence on March 18, 1881. Ella Laurena Lea was proclaimed to be the first Anglo child born in Roswell—a designation that stayed with her all her life.

In the summer of 1882, Sallie Lea took her children and traveled to Mississippi for an extended visit with her family. The operation of the boarding house that was the Lea residence was turned over to young Sophie Alberding. She became acquainted with John Chisum, Sheriff Pat Garrett, and other notables of Lincoln County. One night Captain Lea and Sheriff Garrett called Miss Alberding in for a conference. They told her all about John W. Poe, the sheriff-elect, who was coming to Roswell to recuperate from a fever he contracted in Mexico. Lea told her:

> You see, Sophie, John Poe is going to be our next sheriff, and he's going to be more than that: eventually, he'll be one of the really big men of this whole section. He's a bachelor and—and Pat and I have decided you should marry him. That's what I called you in to tell you.

As outraged as Miss Alberding was at this proposition, she ended up marrying Sheriff Poe at 5:00 a.m. on May 5, 1883, at the Lea house.

Ella Calfee Married Milo Pierce

Ella Lea Calfee began to invest in local projects soon after she arrived in Roswell with her two sons. One of her projects was to join Pat Garrett, Barney Mason, and her brother, J.C. Lea, in the incorporation of the North Spring River Centre Ditch Company on October 1, 1881. She also bought the Bosque Grande headquarters of John Chisum north of Roswell on the Pecos, and then sold it in the fall of 1881 to J. P. White for the beginning of the LFD Ranch.

In October 1881, Milo Pierce, who had been one of the Murphy-Dolan gunfighters during the Lincoln County War, was still on crutches as a result of a gunshot wound he suffered in 1879. One afternoon in what is now downtown Roswell, some cowhands were trying to break a team of horses to pull a wagon. They got the horses harnessed, hitched, and blinded. They then helped Pierce into the seat to drive the team, a feat which "... a whole man might have dodged," according to Mrs. Poe.

Ella Lea Calfee Pierce, Captain J.C. Lea's sister. *Courtesy: HCSNM, Photo No. 488-B*

Milo Pierce, a former gun-fighter for the Murphy-Dolan faction in the Lincoln County War, married Ella Lea Calfee, J.C. Lea's sister. *Photo by Parker, El Paso. Courtesy: HCSNM, Photo No. 3334-A*

When Pierce was ready, the other men removed the blinds. The horses lit out, fighting and resisting their unwelcome assignment, until Pierce and the wagon were out of sight. Before long, here they came back, with the horses trotting along "... as quietly as a pair of old buggy horses."

Watching this performance from a window in Captain Lea's house was Ella Calfee. She burst into tears when she saw what Pierce was about to do, because she feared for his safety. Her tears thus announced her affection for Pierce, and she proceeded to reveal to family and friends that the two of them were planning to be married.

Mrs. Poe indicates that the wedding of Milo Pierce and Ella Lea Calfee was almost a comedy of errors. They set the date and engaged Justice-of-the-Peace Alderberry to perform the ceremony. Alderberry went to Santa Fe on business and became ill while he was there, so he could not make it back to Roswell in time for the scheduled wedding. There was no other justice-of-the-peace within sixty miles, so the happy couple told Captain Lea they had decided to go to Seven Rivers and have the justice-of-the-peace there marry them. Lea hit the ceiling! It just was not proper, even on the frontier, for an unmarried couple to travel together. Lea declared:

> My sister is going to be married right here in my home. You can wait, Milo. And so can you. Neither of you is exactly a young person, but your age does not give you any license to travel to Seven Rivers together—not even to get married!

Resigned to the delay in their plans, the couple awaited Alderberry's return. The wedding was rescheduled for April 20, 1882; and this time it was Pierce who did not show up! He was counting sheep and could not stop until he finished. He finally arrived, and the wedding came off at last. The Pierces had two children: a daughter who died as an infant, and Milo, Jr.

Growth of Roswell

The First Schools. The first school which Roswell children attended was not built in the little village, but three miles east where many children lived on farms. James Miller built the adobe building, and lawyer Asbury Rogers taught a three-month term there in 1881. Five years later the building collapsed, so a new one, called 'The Farms,' was built southeast of town at the location to be known later as 'L.F.D. School.' Sarah Lund opened the school year in the fall of 1886. Miss Lund later married C.D. Bonney, Captain Lea's one-time business partner.

Meanwhile, Roswell residents decided they wanted a school closer to town. They collected funds to build a two-room building near the southwest corner of present East Tilden Street and South Sherman Avenue. Miss Lina Tucker is credited with collecting money from cowboys to help build the school. Called simply the 'village school,' it opened in the fall of 1885 with Fred Farmer as the teacher. The building was used until 1890.

Dunnahoo's Blacksmith Shop. One of the families to settle early in Roswell was that of Rufus H. Dunnahoo. After a short stay in Seven Rivers, the Dunnahoos moved to Roswell in 1881. Dunnahoo soon opened a blacksmith shop on the southwest corner of what later would be Fourth and Main Streets. This was the location where Van C. Smith had built a blacksmith shop about 1870. Apparently, Smith had no permanent blacksmith there, although he may have had temporary workers to come in for short periods. Lea's employees used it as a general shop. Dunnahoo, however, was the first 'permanent' blacksmith and probably used the building put up by Smith. According to Lucius Dills, George T. Davis—who later became Captain Lea's brother in law—bought an interest in the blacksmith shop in 1882. The shop then began buggy, wagon, and wheel repairs. Dunnahoo later moved his shop to South Main Street. He followed that trade for the rest of his working life, but he was also well known in the area as a fiddler and a hunting guide. His son, Rufus J., later became a law officer and was killed in the line of duty in 1931.

Hampton's Store. James H. Hampton put in the second store in Roswell in 1882. Hampton came to the Roswell area in 1878 with the Corn family; Martin V. Corn was his stepfather. Hampton's store was established before the streets were surveyed. After the streets were platted, his business was midway between Third Street and Second Street and on the west side of Main Street. He operated the store until about 1900, according to James D. Shinkle. Hampton was also a bricklayer and built chimneys for many early Roswell houses and businesses.

Blashek's Mill. Another early enterprise was Blashek's Mill, the first industrial plant in Roswell. Captain Lea was very instrumental in attracting the facility to Roswell. Austrian-born George Blashek was a miller in Missouri; in 1881 he sent letters of inquiry to Billings, Montana, and to Roswell. Lea encouraged Blashek to put in the mill, offering to provide land and some financing for it, as well as some help in bringing Blashek's family to Roswell.

Early in 1882, Blashek moved his family to Las Vegas. He then traveled to Roswell to talk to Captain Lea, and they reached an understanding. In October of that year, Lea dispatched one of his men to move Blashek's equipment and belongings to Roswell, arriving back on November 20. The family stayed in Las Vegas temporarily while Blashek worked to start the mill. He suffered a setback when he came down with smallpox, but was able to continue within a few weeks.

On January 13, 1883, Captain Lea deeded eighty acres to Blashek to be used for milling, manufacturing, and agricultural purposes. The agreement required that the mill be built and operational within a year to produce "first-class flour and meal." The land ceded to Blashek is located on the southwest corner of the present intersection of North Atkinson Avenue and East College Boulevard.

The construction of the mill facility began two days after the land was deeded. It was on the north side of North Spring River and about a hundred yards west of present Atkinson. The rock and concrete building also included living quarters for the family. Lea had already taken a ditch out of the North Spring about where the present railroad crosses the river channel; at that time the river channel made a horseshoe bend to the south and back up to the present channel. The ditch carried the water east until it was north of the mill building, at which point it went underground to flow

Above: Blashek's Mill was the first industrial operation in Roswell. George Blashek began operation in 1883 with assistance from Captain J.C. Lea. Photo, c. 1970s, shows some of the ruins.
Courtesy: The Author

Below: Roswell, in 1888, looking north from First Street. The big Jaffa-Prager & Co., Roswell's first department store, is on the right at the corner of Main and Second Streets.
Courtesy: HCSNM, Photo No. 3351

beneath the building, where the turbines were situated, and from there back into the stream. The turbines turned millstones that were on the ground floor. The building was completed in September 1883.

The mill's main function was to custom-grind corn and wheat grown by area residents. At one time, Blashek produced his own brand of flour and sold it throughout the area.

Blashek's family arrived in May 1883. They seemed to be a tragic and peculiar family. Several of their children died violent or untimely deaths. Blashek's sons operated the mill sporadically until World War II, although in the later decades it was mainly an historical curiosity. The mill ruins were completely obliterated in the mid-1980s when Lovelace Park was built, although some vestiges of the concrete-lined ditch still remain.

Death of Sallie Wildy Lea

According to Carole Larson in *Forgotten Frontier: The Story of Southeastern New Mexico* (1993), Mrs. Lea's health declined for several months in 1883-84. Sallie Wildy Lea died on February 20, 1884, at the age of thirty-four years. "The hard, rough life and being far from a physician's care during her last illness was responsible for her early death," Ella Lea Dow said of her mother in 1936. Sallie's death was a tremendous shock for the Captain. Her stunned family buried her in the community burying ground southeast of town. Later called 'South Side Cemetery,' it was located just south of the Río Hondo and a few blocks east of what would later become South Main Street. The cemetery, or much of it, was removed to the present South Park Cemetery location in 1894 to make way for the railroad, according to local historian Georgia Redfield.

At the age of forty-two, Lea was widowed again; and this time he had two small children dependent upon him. Ella was three years old when her mother died, and her father's sister, Ella Lea (Mrs. Milo) Pierce, took her in for a while. Ella Lea Dow described in 1936 how she came to leave: "When four, … my father came one day, and, not wishing to distress my aunt by taking me away, he picked me up without her knowing, and carried

Ella Lea lived with the Thurber family in San José, California, *c.* 1884-89, after her mother died. This picutre was taken during her stay there. *Courtesy: HCSNM*

Below: Sallie Wildy Lea with her two children, Ella and Wildy, *c.* 1883. Photo by Frederichs of Vicksburg.
Courtesy: HCSNM, Photo No. 3340

me to the Thurber's home." There is some evidence that Mrs. Horace Thurber was living at Fort Stanton at the time, but ultimately Ella went to San José, California, with Mrs. Thurber, who also had a daughter. Thurber was an old friend of Lea, a wealthy New York wholesale grocer and partner in the Arbuckle Coffee Company. Ella stayed with the Thurbers for about four years.

Seven-year-old Wildy Lea was also cared for by his Aunt Ella Pierce after his mother died. The Pierces lived in a house that Captain Lea owned at the northwest corner of Fifth and Main Streets, and the Captain also roomed with them.

The property that Mrs. Lea had inherited from her father, including most of the block on which the store and hotel were situated, was passed on to her children. The Captain was in charge of it until they were grown. Their inheritance gave them a source of support for the rest of their lives.

Although he was very mindful of his wife's death, Lea did not write a will until some three years after she died. It was prompted by an accident in early May 1887 when he was returning in his buggy from a trip to Las Vegas. According to later probate testimony by Milo Pierce, Lea's "... team become unmanageable, throwed him out, and come pretty near killing him. He said that he wanted to make a will in case he should happen to an accident like that and get killed, that his children could get his property." Lea's knee was sprained and his side was bruised. A few days after the accident, he asked his brother, Frank, to write up the will. Frank Lea and Pierce both witnessed Captain Lea's signature on the will on May 16, 1887. Lea immediately left in his buggy to visit his cow camp on the Río Feliz south of Roswell, and from there he went on to Texas.

Ada Edwards Davis

Meanwhile, Mrs. Lea's sister, Ada Wildy Edwards, took over the operation of the boarding house, where a sign on the front porch announced that meals were fifty cents. Ada Wildy married William E. Holloman in April 1874 in Mississippi, but he died in October 1875. They had a daughter, Helen Laurena Holloman.

Ada Wildy Holloman, Sallie Lea's sister, as a young woman.
She took over the Lea hotel/boarding house after Mrs. Lea died in 1884.
Ada later married Ernest Edwards and George T. Davis.
Courtesy: Barbara Buckner, Carlsbad, New Mexico

Ada came to Roswell with other members of the family, where she met William D. Edwards, Captain Lea's bookkeeper. They went to Mississippi where she married Edwards in December 1880. After a short sojourn in Roswell they apparently returned to Mississippi for the birth of their son, Ernest, on December 29, 1882. Edwards disappeared from the records by about 1885, so it is not known what happened to him or that marriage, according to Ada's granddaughter Verna Reed of Carlsbad. Ada later married (c. 1888) George T. Davis; they were both about thirty-seven years old. They had four children born to them on their farm near Roswell on what was later to become North Union Avenue. Davis was a blacksmith as well as a farmer. They later lived at Hope in Eddy County.

Sale of Lea, Bonney & Company

Following the death of his beloved wife Sallie, Captain Lea soon left the mercantile business when he and C.D. Bonney sold out in 1884. Bonney established the T Ranch southwest of Roswell. Sarah Lund of White Oaks came to Roswell to teach school in 1885, and it was not long before she and Bonney were married. Bonney later opened 'Bonney & Sons' livery stable in Roswell on the east side of Main Street between Second and Third Streets.

The buyer of Lea's store was William H. Cosgrove. Cosgrove had settled in Roswell in 1878 to help his brother Michael with his mail-carrying contract. When the mail contract expired, Cosgrove went to the East for a year, and then returned in 1884 to buy the Lea & Bonney store.

Dills wrote that when he arrived in Roswell in June 1885 Captain Lea was living in a house he owned at the present northwest corner of Fifth and Main Streets rather than in the boarding house. Sophie Poe describes the house as a brown adobe with a shingle roof. By the time Dills got to town, Cosgrove had taken over the store; he apparently kept the Lea, Bonney & Company name.

Meanwhile, Cosgrove was appointed postmaster of Roswell on April 23, 1884. Four days later he married Sarah Tucker of Missouri. They also lived in the house owned by Captain Lea on the northwest corner of present Fifth and Main Streets where Milo and Ella Pierce later resided. Mrs. Poe states that Cosgrove moved the post office from the store to his house, and he also sold clothing and shoes there. He built a residence for his family at the northeast corner of present Fifth and Main Streets in 1886.

Laying Out Roswell's Streets

The first platting of the town of Roswell in 1885 was one of Captain Lea's most important accomplishments, and was part of the larger plan of his Lea Cattle Company to develop its lands. Dills explained the informal way the townsite was structured before it was platted:

> Main Street was roughly laid out to accommodate these existing structures, and was recognized as a highway, that is, as far as the exigencies required. This Smith & Wilburn corner was taken as a starting point, and the distances north and south along the street were stepped off or otherwise guessed at; and this guessing accounts for the fact that when the town was finally laid out and platted, it was found necessary to make the tiers of blocks between Second and Fifth Streets odd sizes, to avoid interfering with existing buildings.

Dills goes on to describe how the town was officially platted:

> By 1885, farms had been opened up, irrigation ditches constructed, and cottonwood trees planted along them on the greater part of the lands within the sector; and Roswell had grown into a hamlet of ten houses—nine on the west and one on the east side of Main Street.
>
> Having reached these proportions, Captain Lea had his brother, Alfred E. Lea, to come down from Denver and definitely lay out, into blocks and lots, what was then considered to be adequate for the near future growth of the hamlet. This work was executed faithfully and efficiently, and the plat thereof was certified as finished by A. E. Lea as of November 1, 1885 ... .

The city plat was officially filed in 1887 with the Lincoln County clerk, about two years after it was drawn. An examination of the original document in the Chaves County clerk's office in Roswell reveals that some additions and changes had been made

Alfred E. Lea, brother of J.C. and Frank, surveyed the Roswell townsite and drew the first plat of the town (see below) in 1884.
Courtesy: HCSNM

Detail of Plat of Original Roswell, 1885. The large shaded area in Block 4 apparently was a planned building that was never built.
Courtesy: Original in Chaves County Clerk's Office, Roswell.

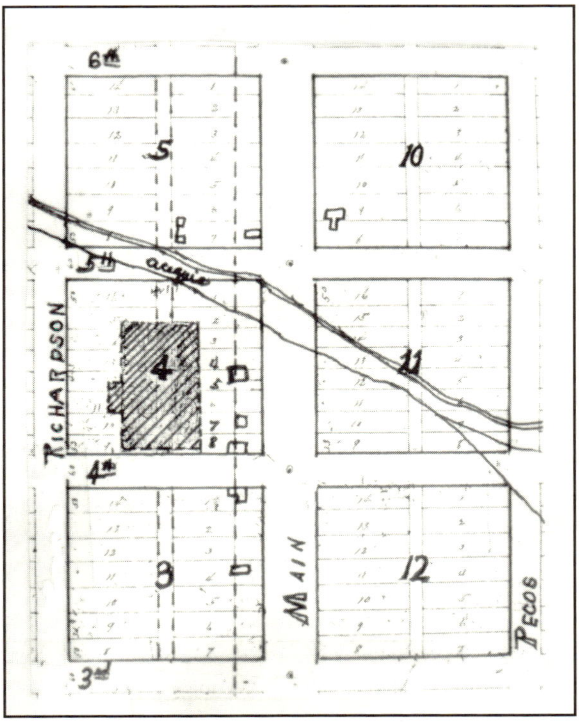

to the plat between the time Alfred Lea completed it and the time it was filed (and probably afterwards), including some buildings (or proposed buildings) and the name of Richardson Avenue (originally Overton, then Capitan).

The area platted ran from First Street north to Eighth Street and one block east and west of Main Street. Most of the city blocks in the original townsite as surveyed by Alfred Lea are about 350 feet long, with seven lots facing the north-south streets. However, the 400-block of North Main is about 384 feet long and has eight lots on each side of the alley (Block 4 of the original townsite of Roswell) because of existing buildings at the time of platting. The extra length of the block appears to accommodate the width of the Smith-Lea store building at the south end of the block. Allowing for existing buildings also seems to account for a slight disorientation of Main Street from the compass.

Poe, Lea, & Cosgrove

The successor to Lea, Bonney & Company took on new dimensions when William H. Cosgrove took in two partners and expanded his business in the late 1880's. Cosgrove's partners were John W. Poe, the famous former deputy and sheriff of Lincoln County; and Jesse Smith Lea, J.C. Lea's cousin, later mayor of Roswell and holder of several state and county offices. These men already had a real estate company—Poe, Lea, & Cosgrove—before they bought the mercantile, which, deed records show, bought the Lea & Bonney site on December 1, 1886. Carole Larson avers that the partnership was not established until 1888.

Poe, Lea, & Cosgrove made extensive changes to the Smith-Lea store building. They built a "false front" on it with two entries, which was a twenty-two-foot eastward extension of the original building out to the present frontage line. This made the store about thirty-by-seventy feet, outside measurements.

By 1889, Poe, Lea & Cosgrove had completed a second major remodeling of the building. They made the structure taller and added a new porch. A sign on the front, according to a photograph, reads, "Poe, Lea & Cosgrove - Dealers in General Merchandise."

J. Smith Lea, cousin to J.C. and Frank, January 18, 1906, at the age of 50. Smith Lea was the 'Lea' in 'Poe, Lea & Cosgrove.' He later served as mayor of Roswell and held several state and county offices.
Courtesy: HCSNM, Photo No. 3333

William H. Cosgrove bought out Lea, Bonney & Co., in 1884, and later expanded to 'Poe, Lea, & Cosgrove,' until he was appointed receiver in the U.S. Land Office in 1893. Photo *c*. 1910.
Courtesy: HCSNM, Photo No. 492

It seems likely the new owners retained most of the walls of the original Van C. Smith structure and added on to them.

One source states that Poe, Lea & Cosgrove operated for five years; another claims the store was sold in 1892. An advertisement in the January 15, 1892, issue of the *Roswell Record* offers "5 Gal. Coal Oil - $1.75; 50 lbs. P. D. Flour - $1.85; Coarse and Fine Salt a Specialty - until we are closed out. Call and ask for prices on everything we have." This ad suggests that they were closing out the stock and not just selling the store.

There is some confusion about what Poe, Lea & Cosgrove did with the original Smith-Lea store building. An article in the April 8, 1892, *Roswell Record* states that the original store building was "... now used as a warehouse by Poe, Lea & Cosgrove ... ." An item in the April 27, 1900, issue of the *Roswell Register* states the original store building was *north* of Poe, Lea & Cosgrove's store. If correct, this means that their main store was to the south, in the 300-block. There does not seem to be any evidence in the deed records that they bought any lots in the 300-block, but it is possible that they leased or rented a building there. Deed records show that Poe, Lea & Cosgrove sold the lots and building of the original Smith & Wilburn location to their own real estate company on April 30, 1896. Cosgrove resigned as postmaster on May 22, 1893, two days after learning that he would be appointed receiver in the U.S. Land Office in Roswell, a position he held until 1898.

8

THE 'LAND AND CATTLE' BUSINESS

J.C. Lea's Ranching Operations

The cattle operations of Captain J.C. Lea were representative of hundreds of cattlemen all over the West. Lea's activities were typical of the Cattleman's Frontier as the cattle industry evolved from open-range grazing and long trail drives to settled cattle raising on leased or owned—and sometimes fenced—land. In southeastern New Mexico Territory, sheep often were as important or more so than cattle. Lea was one of the first graziers to run cattle and sheep successfully on the same range.

When the Leas moved from Colfax County to Lincoln County in the fall of 1877, they brought a good-sized flock of sheep and a small herd of cattle. In addition to filing on the land just west of the hotel and store in Roswell, in 1879 Lea began to acquire large amounts of rangeland.

The federal government program which made it possible for Lea to buy public lands started in 1870, during a two-week auction of lands along the Canadian and Hondo Rivers. Some 1,958 acres were sold for about $2,500; but no bidder would offer more than the $1.25 per acre going rate for pre-emption land. After the period during which the land was auctioned, it was opened for 'private filing.' This meant a buyer could buy the land from the federal government at a set price. One of the largest purchasers during the sale was Wilson Waddingham, who bought much land along the Canadian River and Ute Creek near their confluence that was developed into the famous Bell Ranch in the Tucumcari area.

Victor Westphall, in an article in *New Mexico Historical Review* (January 1958), states:

> … extensive purchases at private entry were made by Joseph C. Lea along the Río Hondo from its source to its junction with the Pecos River. From 1879 through 1885, he bought 13,386.98 acres. Other members of the Lea family bought more than 2,400 acres in the same area.

Lincoln County deed records indicate that one major purchase was the ranch of Saturnino Baca, prominent Hispanic leader in the county. Chaves County deed records show that a dozen of the patents secured by Lea were for tracts comprising over 1,200 acres within the future town site of Roswell. Lea had a very active ranch almost from the beginning and employed numerous herders and other workers.

In the summer of 1881, Captain Lea sold some of his sheep and concentrated on raising cattle. James M. Miller and his brother Bill had settled near John Chisum's South Spring River Ranch southeast of Roswell in an area that has long been known as Oasis Ranch. Miller's cattle did not thrive in that location, and Lea—for reasons known only to him—offered to trade some of his sheep for the cattle. Lea sought Chisum's approval before proposing the deal to Miller. It took some convincing to get Miller to agree to become a sheepman surrounded by cattlemen, but he did agree.

The sheep could not be delivered until the lambs were weaned, which would be around the first of October. Miller described the event in his autobiography, an unpublished manuscript in the archives of the Historical Center for Southeast New Mexico in Roswell:

> In those days we never heard of such a thing as a chute to help in separating sheep. We simply caught them and dragged them out from the flock. This is the way we did it when I went to Captain Lea's to get the sheep I was to have in exchange for my cattle. His entire herd of 5,500 sheep was first put in one large corral. Then we caught and dragged out the wethers, something over 2,000. Next we did the same with the lamb crop, and put them in another corral. Then we counted out 250 ewes, taking them as they came, all sizes and ages. In like manner, we counted out 250 lambs. When I had gotten all my sheep,

> Captain Lea's herder, Jeff by name, turned all the rest of the sheep out of the corrals and drove the whole herd of 5,000 to his camp.
>
> As I could not drive my small herd that night to my farm, I placed them in the large corral that then stood just west of Captain Lea's residence, intending to drive them in the morning. The corral was a very old one and the adobe in it so mixed with alkali that the walls, some 7 feet high, were getting crumbly about the foundation. …

Miller spent the night at Lea's house in Roswell; after he had gone to bed, Lea came to Miller's room to tell him that the entire north wall of the corral had collapsed and fallen on his sheep. Five were killed and two died later; the rest got out of the pen. It was not as bad as predicted, however, as they had little trouble rounding up the flock the next morning. Miller and his workers drove the flock to his place that day and arrived with 493 sheep to begin his career as one of the leading sheepmen of New Mexico. Lea continued to raise sheep as well as cattle for many years.

Although J.C. Lea had vast land holdings and thousands of cattle, he wanted to expand his operation. In 1883, he organized a co-partnership with several other people, none of whom lived in Lincoln County or even in New Mexico. This was the first 'Lea Cattle Company.' The other partners were Wallace N. March, Augustus H. Powell, Seneca D. Powell, J.E. Powell, W.H. Colbern, W.A. Noel, James Noel, and George W. Scott. It is not known where any of the partners lived except Scott, who lived in Jackson County, Missouri—Lea's old home county. The other partners furnished the finances, while Lea earned his twenty percent interest by working as general manager for the company. Few records of this company are known to exist.

Lea Cattle Company

Two years after forming the co-partnership, Lea set his sights on still bigger stakes. According to an account in the *Roswell Record* of February 12, 1904, "… he was engaged in the cattle business and when the crash of '84-'85 came, like thousands of others in the Southwest he was caught in the whirlpool and left with a load that few men would have attempted to carry." Another view of his motives is expressed by Matt Dodge in the May 1978 issue of *Real West:* "Captain Joseph Lea … had some grandiose ideas and wanted to emulate his neighbor, John Chisum He cast about for a money man and hooked Horace [Thurber], who had all the symptoms of the cattle-fever syndrome."

Horace K. Thurber was a wealthy partner in the United States Steamship Lines and the Thurber-Arbuckle wholesale grocery firm of New York City with John and Charles Arbuckle, manufacturers of the famous Arbuckle Coffee. Like so many Eastern businessmen, Thurber was fascinated by the idea of investing in Western cattle ranches. Apparently envious of the land holdings of John Chisum, C.C. Slaughter, and Wilson Waddingham, according to Gene M. Gressley in *Bankers and Cattlemen* (1966), Thurber bought several sections of land in Lincoln County. He had visited the Fort Stanton area as early as 1883. Together with John C. DeLany, post trader at Fort Stanton and former partner of Will Dowlin in Lincoln, Thurber incorporated the El Capitan Land and Cattle Company on April 23, 1885, for $300,000 in paid-up capital.

Lea contacted Thurber and presented a very optimistic proposal for big-time cattle ranching at the same time that Thurber and DeLany were forming their company. Gressley states the Arbuckle brothers joined Thurber in the PO Ranch near Cheyenne and in the Lea Cattle Company of Lincoln County, New Mexico. They no doubt were stockholders but were not in on the formation of the company.

The articles of incorporation of the Lea Cattle Company are included in the book of minutes for the board of directors; the book is in the possession of Mary Herrington of Roswell. The articles include only the names of Horace Thurber, Lea, DeLany, and Charles S. Thurber, Horace's brother, "… a majority of whom are … residents of the Territory of New Mexico." The document, dated April 20, 1885, to be effective on May 1, declares, "… we do hereby associate ourselves together and form a body politic and corporate under the name and style of 'LEA CATTLE COMPANY.'"

Above: James M. Miller (r.) with James M. Mullens. Miller bought sheep from Captain J.C. Lea and became a leading sheepman of New Mexico. *Courtesy: HCSNM, Photo No. 3167*
Below: Lea Cattle Company Share No. 128, issued July 1, 1885, signed by H.K. Thurber, president, and H. Aplington, Secretary-Treasurer: *Courtesy: HCSNM, Archival Collection 980109*

UNITED STATES OF AMERICA.

LEA CATTLE COMPANY.

No 128 No 128

LINCOLN COUNTY, NEW MEXICO.

1000

Know all men by these presents; that this Company a body corporate by virtue of the Laws of New Mexico, do for value received promise to pay to or bearer, the sum

ONE THOUSAND DOLLARS

in gold coin or in legal tender currency, on the 1st day of July 1900, at the Importers and Traders National Bank in the City of New York, with interest at the rate of six per cent per annum payable semi annually in gold coin or legal tender money of the United States on the presentation and surrender of the coupons or interest warrants annexed, at the Importers and Traders National Bank in the City of New York. This Company reserves the right at any time after five years to pay the principal of this Bond, on the first day of July at the time the coupons become due, by giving notice to the holder of the Bond on the first day of June of their intention so to do. This Bond to be valid must bear the signature of the President and Secretary and be countersigned by the Treasurer.

Dated Fort Stanton Lincoln County New Mexico July first, 1885

SECRETARY. PRESIDENT.

COUNTERSIGNED BY TREASURER.

The purposes of the corporation were "… breeding, raising, buying and selling cattle and other domestic animals, and buying and selling merchandise." The capital stock was $1,000,000 in ten thousand shares; and the life of the corporation was to be forty-nine years. Lea went to New York to close the deal; Charles Thurber, who claimed to live in Lincoln County, came a few weeks later to Roswell to sign the document.

The corporation designated Fort Stanton as the location of the company headquarters. The board of directors had an inauspicious start, being unable to reach a quorum on June 1 and June 6, 1885, when only Lea and DeLany showed up at Fort Stanton. Finally, on the evening of June 15, Charles Thurber appeared. With Captain Lea presiding, they elected H.K. Thurber as both president and treasurer, with H. Aplington of New York as secretary in order to avoid having to send papers to New Mexico for signatures. When the meeting resumed the following morning, the directors voted to issue $250,000 in company bonds to buy land and cattle in accordance with the agreement Horace Thurber had reached with the 'old' Lea Cattle Company. They also voted to give Thurber another quarter-million in bonds and nine thousand shares. The other thousand shares were issued to Lea, who was also named ranch manager. Later, Lea became president and Thurber became secretary-treasurer. According to Padgitt, Thurber invested $500,000 in cattle and another $250,000 to buy additional land with water rights. On June 26, 1885, the old Lea Cattle Company sold 3,364.3 acres of ranchland to the new Lea Cattle Company for $250,000, according to Chaves County deed records. The company record book also shows that the new company bought 105 horses and eleven mules from the old company. Lea sold nine thousand cattle to the company for a quarter-million dollars. Thurber's money was used for the operation of the ranch as well as buying land and water rights. The company paid Lea $90,000 on July 3, 1885, for 1,440 acres of his land, the largest of several such transactions. Lea transferred so much of his land holdings to the company that it seems impossible to separate his interests from those of the corporation.

At a called meeting on July 5, Horace Thurber was elected president, Charles Thurber was designated vice-president, Lea was named ranch manager, and DeLany was chosen as secretary. Horace Thurber gave up the position of treasurer, and Henry Aplington was named to replace him. W.A. Parshall of New York was designated as temporary assistant secretary.

The expansion of Lea Cattle Company was quite successful. On September 7, 1885, the *Las Vegas Daily Optic* stated, "Captain Lea controls fifty miles of river frontage and numerous springs in Lincoln County." The letterheads and bonds of the company picture cattle and horses bearing the 'L-E-A' and 'H-K-T' brands. The letterhead defines the company range as "Agua Azul, Rio Hondo, and Rio Pecos, Lincoln County." They called the ranch 'Agua Azul Ranch,' after a pool where one Jesús Maes had settled in 1878. The ranch headquarters were at the eastern foot of the Capitán Mountains. Padgitt avers that the ranch ran 30,000 to 50,000 head of cattle. George Curry (*George Curry 1861-1947: An Autobiography,* 1958), who later was governor, states that after he was elected assessor of Lincoln County in 1890, he surveyed the land records and traveled widely throughout the giant county. He found that the Lea Cattle Company ranch, with 40,000 cattle, was the largest cattle owner in the county and probably in the Territory. The *Illustrated History of New Mexico* (1895) asserts that the company owned over 22,000 acres of land. They also used the open range public lands, where their cattle ran more-or-less together with everybody else's cattle.

For more than a year, Lea spent Thurber's money as if it were going out of style. He bought cattle, built irrigating canals, raised grain, and had the town of Roswell platted for future development. He turned his empire into what the *Daily Optic* called "a Garden of Eden in what was already paradise."

Trouble on the Range

Much of the early sheep-raising in Chaves County was carried on by itinerant sheepmen who caused conflict with cattle ranchers in the 1880s by grazing their sheep on the public domain and watering them at private sources. As a result of overgrazing, a law was passed that required the flocks to move at least six miles per day. This helped some; but uncooperative sheepmen would graze out three miles on one side of a stream, cross over and graze out three miles on the other side—thus obeying the letter of the law.

At the time Lea and Thurber were getting their cattle company off the ground, Lea learned that someone had reported to Governor Edmund G. Ross that a state of war existed between the cattlemen and the sheepmen in Lincoln County. Lea took exception to this allegation and penned a lengthy letter (Ritch Collection, State Records Center and Archives) to the governor—in which he identifies himself with the sheepmen—from Agua Azul Ranch on July 8, 1885:

> … I must respectfully beg leave to State to you that such is not the case & that a great injustice has been done to both cattle & Sheep men by conveying to you Such false impressions. There never was a time in the history of Lincoln County when so friendly a feeling existed between the cattle & sheep owners as at present, with the exception of about three owners of Sheep herds who have for the last three to twelve months greatly imposed upon Some of the cowmen who are & have been among the most worthy & law abiding citizens of the county. Their actions throughout has been as well as their talk that of bulldozing & blackmailing certain cow men, whose characters are & have always been unimpeachable & among the most useful & best citizens of the country.
>
> I have owned sheep in Lincoln County for the last 8 or 9 years & still do, but have never had the least trouble with cow men & must say from experience that no trouble is likely to occur unless cow men are grossly & repeatedly imposed upon by the Sheep. The cow men as a rule are now the wealthy & most influential of our county & cannot afford to encourage lawlessness or allow it to go on ... .
>
> It is true that within the last year or 18 months many violations of law have occurred but we consider this no fault whatsoever of the officers. This county being nearly 200 miles Square It is impossible for an officer to be Stationed as a guard in so many different localities. This is a border county & has always been a receptacle for hard cases & for the last 18 months an increased number has been coming.

I am requested by many good citizens to ask you to come to the county in person & get acquainted with the people before making up your mind ... .

I write you at the request of good men of the county & in the interest of law & order. Should you listen to some letters you might get our county may be thrown into a worse state of anarchy than it was in the days of the Lincoln Co. war & Billy the Kid &c. As it is I think we will soon hear no more of the disgraceful conduct that has brought our county in disrespect.

We feel sure that we have a Sheriff who will do his duty – in whose capacity & honor we can fully rely & that our troubles have been no fault of his ... .

Trusting that we may see you Sometime among us & that our local disturbances will soon be a thing of the past.

I am, With Much Esteem, Your Most Ob't. Sv't.

J.C.Lea

Roswell NM

Thurber Unhappy with Lea's Management

The cattle country experienced summer drouths and winter blizzards from 1885 to 1887. The bottom fell out of the cattle market, and only those with large holdings and plenty of cash reserves were able to survive. Lea apparently continued to build his cattle empire even in the face of low cattle prices. In late 1886, Thurber began to complain to Lea about Lea's free-spending ways and lamenting that he was not seeing a return on his investment. On December 15, he wrote to Lea (cited by Gressley):

> Those cattle notes have got to be paid when due, and the heavy drafts you have been making on me have, as I fully explained to you before, taken all the money that I am willing to invest in Lincoln County, and I want to begin to get some of it back at as early a date as is possible.

Thurber's position irritated Lea and he offered to resign, but he made no apparent effort to change his methods. The financier knew he could not operate the ranch from New York, so he eased the pressure on Lea for a couple of years.

Many cattlemen had trouble raising the needed cash flow for their operations, according to Gressley. The investors learned, to their dismay, that the initial capitalization of a ranch usually did not last more than one year. Friends and relatives of the cattlemen typically would be called upon for funds, and they usually responded. Many of those friends and relatives lost their enthusiasm for investing in cattle ranches after they realized no dividends were forthcoming, loans were in default, and interest payments were behind schedule. Lea expressed this reality in a letter to John DeLany on May 9, 1887, having recently returned from the East (cited by Gressley):

> New York was a more friendly place three years ago. Now all I hear back there is questions as to what we are doing with our money, why the calf crop isn't better and when will we again be able to pay our notes. As you know, I am afraid my answer didn't satisfy their questions.

Gressley maintains that the alliance between Thurber and Lea was quite representative of "cattlemen's politicking." Thurber had many years of experience and success as a Wall Street investor, and he gave a lot of advice to Lea about Western politics and the economics of cattle raising. Thurber was an Eastern conservative Republican who may have had some misgivings about Lea's Democrat position. He tried to convince Lea that Republican principles were superior; and even if Lea disagreed with the Republican principles he should take up their platform because they were the party in power.

When Lea contemplated how he might get a railroad to come to his property in 1888, Thurber sent him detailed advice on how to pass such a bill through the Legislature. Thurber saw the possibility of a railroad as a way to recoup his losses on the ranch. On December 20, 1888, he wrote to Lea (cited by Gressley):

> I do really hope that the town property will make a great deal of money, but Captain, as soon as I could sell off and while the excitement is on, I would sell, so as to reduce our holdings.

> The time to sell in a town is when the railroad is building; the moment it goes through [,] the moment the excitement dies away [,] some other town comes up in its stead. It would be a great thing, if you could get the repair works and the car works at your town."

It was to be several years before the Pecos Valley Railway reached Roswell from Eddy. Until it came in from the south, Roswell claimed to be farther from a railroad than any other town in the country.

The minutes book of the Lea Cattle Company is silent for the period from July 1885 to April 1888. Probably the minutes were kept elsewhere and have not survived with the book. The record for April 10, 1888, indicates that the regular annual meeting of the stockholders took place every April at Fort Stanton. Lea, DeLany, and Charles Thurber were present for the 1888 meeting. The only business was to accept the resignation of DeLany as secretary and board member of the corporation.

The absent Horace Thurber was named acting secretary when the board next met on April 24 in Roswell for the 'annual meeting' of the directors. This time Lea, his brother Frank, and Charles Thurber were present as directors. They voted to change the corporation's "principal office and place of business" to Roswell from Fort Stanton. Lea established the office of the company at the present 111 West Fifth Street, on the northeast corner of Fifth and North Richardson Avenue. Horace Thurber was elected president and treasurer, Charles Thurber was elected vice-president and secretary, and J.C. Lea was named manager. The board issued an invitation to Horace Thurber to "… make us a visit at as early day as is practicable to look over the property and affairs of the Company." No evidence has surfaced to indicate that H.K. Thurber ever visited Roswell after the cattle company was organized.

Despite the rancor between them, Lea was not hesitant in seeking Thurber's help to secure political favors, Gressley claims. When there was some question about whether or not William H. Cosgrove would continue as postmaster of Roswell in 1889, Lea asked for Thurber's help. Thurber wrote to Lea on June 12, 1889,

"You may rest easy there at Roswell, Mr. Cosgrove will remain postmaster. I have it all fixed and if there is any other place where having a postmaster will do the Republican party any good, I can fix it, if I know about it soon." Lea promptly sent him the names of six other communities.

Lea explained to Thurber in a letter on October 25, 1888, that his desire was "… making this the best ranch in New Mexico." In view of the amount of money Thurber had invested without seeing any profit, it is easy to understand why Thurber had a problem with Lea's ambition. However, his letter to Lea on December 21, 1888, reflects his acquiescence to Lea's management style: "Captain, I have the greatest confidence in you—in your energy, honesty and earnestness, and I wish it were in my power to do all you ask of me."

Just days after Thurber wrote that letter, a new batch of bills arrived from Lea. Thurber wrote another letter on December 29, which was much more strident that the one he had written eight days earlier: "Won't you please tell me the cost you have in regard to your cattle ... ? The reason I want this is, I wish to compare your expenses with those of other ranches, I have met so many people here who are in my situation."

Gressley states that Thurber's forthright strategy may have been designed to alarm Lea and compel him to strive for the benefits of cost-effective management, but it did not work. "A more tempestuous exchange than the Thurber-Lea correspondence of 1888-9," Gressley writes, "would be difficult to imagine." Thurber had seen his investments of hundreds of thousands of dollars in New Mexico Territory pay off in only two of the years he had been investing.

Thurber told Lea in a letter on May 11, 1889 (cited by Gressley), that he wanted very much to see a twelve percent return on his Lea Cattle Company investment; but by that time the company was in considerable distress. A petition was presented to the board calling for a special meeting to be held the day prior to the scheduled annual meeting, which was April 16, 1889. Such a meeting was conducted "... for the purpose of devising ways and means to enable the board of directors ... to bargain, sell, and convey real estate belonging to said company ... ." The record shows that Lea now had 2,500 shares, plus proxy for Horace Thurber's 4,700 shares. John W. Poe and J. Smith Lea had a hundred shares each, and Poe apparently was behind the calling of the special meeting. They proceeded to "empower and authorize" the president to carry on with land sales.

The directors met for their regular annual meeting "... at the rooms of Joseph C. Lea in Roswell ... ." J.C. Lea, Smith Lea, and John Poe elected themselves plus Horace Thurber to be directors for the ensuing year. The meeting may have been somewhat tumultuous, as the minutes book has four versions of the minutes. The first two are in J.C. Lea's handwriting but are "not signed as the entry is erroneous." The official versions are in J. Smith Lea's handwriting; he was elected secretary of the company. J.C. Lea was elected president, and H.K. Thurber was chosen treasurer. New by-laws were adopted to make the land sales legal that were approved in the previous day's special meeting, but it was to be some time before land sales would help the company's cash flow. The minutes book has no records from April 1889 till October 1892.

Thurber was fit to be tied as more and more of Lea's drafts arrived at Thurber's New York office. In the summer of 1889, he began to lay down the law to Lea. He wrote on June 19 (cited by Gressley):

> Now, my dear Captain, it is simply impossible; I cannot stand your drafts. I have told you over and over again that I have got to limit where I could go. I have told you that it was not a question of desire, but it was a question of necessity. I have advanced there to you about $800.00 in cash, and it simply is impossible to me to go any further. You tell me that a portion of this work is to be charged to you; suppose it is, that don't help me a bit. There is a large amount of back interest due me from you which I have not asked for, because I knew you could not pay it.
>
> Captain, you have gone against my wishes, and simply draw on me, and I have paid and paid until I can go no further; and

> now I must request you peremptorily to stop all improvements and stop at once, unless you want your drafts returned to you and you be obliged to raise the money there at one enormous rate of interest....
>
> Now my dear Captain, don't think that this is pleasant for me to say to you, but it is necessary for you will certainly have to get along on your own resources there. I must confess to my great disappointment at your going ahead as you have, when I told you for months past that there was a limit beyond which I could not go.

Thurber kept up the pressure in another letter (cited by Gressley) to Lea dated August 27, 1889:

> Now, Captain, I want to reiterate, and I want you to believe that I am fully in accord with you and I firmly believe that you have done the best you could. The very large amount of money that I have put down there and so much that I didn't intend to, has put me in such shape that I can't get away. I have simply got to stay here and keep things going so as to pay the money that I have had to pay on those ranches.

Lea's unwillingness or inability to comply with Thurber's demands for economy in the operation of the ranch is never explained. Gressley speculates that Lea was committed beyond the point of no return in his efforts to develop the company's property and could not turn back; or, perhaps his disposition was such that he could not refrain from spending someone else's money. Gressley implies that Lea did not understand the big picture of the range cattle business and was unaware of the risks involved. With Lea's proclivity for spending, "Thurber was foredoomed to financial disaster." Strangely, Lea's family seemed to be unaware of his deep involvement with Thurber and his financial problems.

One of Lea's herds encountered problems on a trail drive to Amarillo in October 1889. His men were trailing a herd of eight hundred cows and calves, and they had to cross XIT Ranch lands to reach Amarillo. The ramrod and four of the drovers were arrested for trespassing. According to newspaper accounts, the XIT was trying to "build up Texline at the expense of other towns."

When the Lea men headed for Amarillo instead of Texline, they incurred the wrath of the XIT.

Lea Cattle Company's Land Development Program

Following the formation of the Lea Cattle Company in 1885, Lea continued his development of irrigation canals—sometimes in his own name, sometimes in the name of the company, sometimes both. Selling off ranch land to be settled and cultivated by farmers was a typical practice among 'land and cattle' companies in the West. Lea and especially Thurber hoped the land promotion would turn the Lea Cattle Company into a profitable venture. As a part of the broader plan of attracting settlers and businesses, Lea had his brother Alfred survey and plat the town of Roswell in 1885. Having the streets and alleys laid out made for a more coherent plan of future development.

The land and ditch transactions carried out by Lea make up a perplexing maze of legal documents; the index alone of Lea's deals listed in the deed records in the Chaves County clerk's office in Roswell runs to more than fifty pages. Documents and maps on file at the office of the Pecos Valley Artesian Conservancy District in Roswell also provide much useful information on the ditches.

According to attorney Bernice D'Abadie, expert on the legal aspects of the early ditches of Chaves County, the Lea Cattle Company and Patrick H. Boone incorporated the Hondo Falls Ditch Company on November 3, 1885. The ditch, also known as the Russell Ditch, was located about six miles due east of the townsite. They built a rock dam on the Hondo a mile west of the Pecos River, below the confluence of the North Spring River and the Río Berrendo with the Río Hondo. They soon sold the ditch company, but Lea was again involved with the company in the 1890's. Boone, C. D. Bonney, Asbury Whetstone, and others were in on the early efforts in the same area to build what remains in operation until the present time as the Northern Canal (or Hagerman Canal).

A rather small but very important project was undertaken by Lea in 1886 to provide water for the town of Roswell. This water supply ditch was usually called the 'Town Ditch,' but it was also known as the 'J.C. Lea Ditch.' Lea built a dam at a small spring a few hundred yards above the Lea-Cunningham dam on the North Spring River and ran a ditch about two miles eastward. Early photographs of Roswell suggest this ditch was a continuation or re-opening of the original ditch found by Fellows in 1866-67 and used by Van Smith in the 1870s. The ditch traversed the lots north of Captain Lea's residence near present Fifth Street, crossed what was to become Main Street, proceeded east across the present courthouse square and on toward the farms east of town. Bridges across the ditch were simply two-by-twelve planks. After the town was platted and the courthouse was built, the ditch likely was straightened and re-channeled along the north side of Fifth Street. It fed into the Pierce, Cunningham, & Ballard (Centre) Ditch in the 200-block of East Fifth Street. Men with wagons and barrels peddled the water to residents for twenty-five cents a barrel. The Town Ditch; the Pierce, Cunningham, & Ballard (Centre) Ditch; and the Lea-Cunningham Ditch all continued to flow across downtown Roswell well into the twentieth century.

One of Lea's most ambitious projects was the Lea-Cockrell Ditch. J.C. and Frank Lea, Ada Edwards, the Lea Cattle Company, and John J. Cockrell formed a corporation on June 12, 1887, to take water out of the Río Hondo. The dam was about 4½ miles southwest of the townsite. The ditch is described in the articles of incorporation as being taken out of the south bank and was to be ten feet wide and at least two feet deep. It took an easterly direction and had an approximate length of five miles.

A statement in the articles of incorporation of the Lea-Cockrell Ditch Company gives some indication about the planned direction of the Lea Cattle Company: "The further object and purpose of said corporation being the improvement of lands in connection with said ditch and the colonization thereof and the acquisition and sale of lands under said ditch." The directors of the company obviously wanted to sell much of their land to farmers, and they hoped to attract settlers by the development of irrigation systems in the semi-arid area of Roswell. Actually, according to maps prepared for the Pecos Valley Artesian Conservancy District by

Above: J.C. Lea and his many partners, as well as numerous others, built an elaborate system of irrigation ditches such as this one near Roswell, *c.* 1890s. By Frost's Views. *Courtesy: HCSNM, Photo No. 1814*
Below: Many of the irrigation canals taken out of the North Spring River crossed over the Río Hondo at some point. Because the original flumes were wooden, they leaked and eventually had to be replaced. In this photo, L.B. Craig & Co. is building a concrete flume over one of the Hondo crossings. Photo by Wilkinson. *Courtesy: HCSNM, Photo No. 3445*

Stanley Hordes, the Lea-Cockrell Ditch Company eventually had two ditches coming from their dam and numerous lateral ditches which carried water to several sections of land located between present Roswell and the Roswell Industrial Air Center along both sides of present South Sunset Avenue. Some of the ditches went as far east as the land along present South Atkinson Avenue between East Poe Street and East McGaffey Street. Records indicate the ditches irrigated more than 7,120 acres from 1888 to 1892. The construction of this vast network of irrigation ditches was a huge project undertaken by Lea with Thurber's money.

Although the Hondo ran full for the first year or two of the Lea-Cockrell operation, it was not always as reliable a source of water as the North Spring and South Spring because it depended mainly on runoff. Farmers in the mountain valleys to the west used most of the water, so in some years the volume of water coming down the Hondo was insufficient to irrigate the amount of land under the ditches. The land under the Lea-Cockrell ditches was essentially the same land that farmers attempted to water from the Hondo Reservoir from 1907 to 1916. There was not enough water then, either.

A ditch that used water from the South Spring River was that of the South Spring River Ditch and Irrigation Company. There are a few documents on file at the Pecos Valley Artesian Conservancy District, but no records have surfaced to indicate when and by whom the company was organized nor exactly where their ditch was located. The records indicate that the ditch was used to water 150 acres in 1889, apparently by Rufus H. Dunnahoo, who filed a desert claim on December 16, 1889. However, the records show that the water priority was not established until 1891. Lea Cattle Company sold ten shares of its interest in this ditch to Dunnahoo in March 1891, described as "ten miles southeast of Roswell." The company bought a twenty percent interest in the ditch from W.M. Atkinson in January 1893. Ultimately, the cattle company acquired all of the South Spring River Ditch.

The ownership of the various ditch corporations is a very complex issue. Everyone who owned or rented land 'under the ditch' wanted a portion of the water. Shares—and fractions of shares—of stock in the companies changed hands often, as did the majority ownership of the corporations. Lea and his numerous relatives and partners were involved in most of the ditches in the Roswell area, as were many other people. For example, Lea and C.D. Bonney bought into the Pioneer Ditch in 1886. J.M. Harris, A.H. Whetstone, and a dozen others incorporated the Pioneer Ditch Company of Lincoln County on November 24, 1880. Their ditch, which started carrying water from the south bank of the North Spring River in 1881, began about a mile northeast of town and ran some four miles eastward.

Lea's grandiose plans for taking on huge, costly projects for the cattle company attracted national attention. The *Field and Farm* newspaper, published in Denver, had an article in its January 5, 1889, issue about the Lea Cattle Company's ditch construction project. The paper reported that late in 1888, Lea had a giant V-shaped plow custom built (Gressley says it was made in Denver). "They work fourteen head of cattle on it, and it moves an immense amount of dirt in a day," the newspaper states. "They claim to have the largest team in New Mexico on their work." The ditch was finished around February 1, 1889, and was called the Lea Cattle Company Ditch. Located some six to seven miles southeast of Roswell, it took water from the South Spring River. The course of the ditch was a straight line for about a mile or so southeastward from the river. Records at the Pecos Valley Artesian Conservancy District show the ditch irrigated three sections of land in 1889.

Field and Farm states:

> This is a novel way for a cattle company to operate, but it goes to show that the rangers are awake as well as the grangers, and the percolating waters diverted from the streams will work untold wonders in the industries of the country.

Projects of this magnitude were too costly for Thurber's comfort.

Another article in *Field and Farm* in its November 30, 1889, issue describes a six-mile flume which Gressley asserts was to carry water to "... the desert on the Lea ranch." The cost of the

Above: Roswell was a cowtown for many years, as shown by this herd of Hereford cattle moving south on Main Street, *c.* 1900.
Courtesy: HCSNM, Photo No. 514

Below: The blended family of Captain J.C. Lea and Mabel Doss Day Lea. l. to r.: Captain Lea, Ella Lea, Willie Day, Wildy Lea, and Mabel Doss Day Lea. January 1, 1891. *Courtesy: HCSNM, Photo No. 651*

project was reportedly close to $50,000, Gressley claims; and the land was to be planted in alfalfa. Actually, the article in question is about Thurber's other ranch, El Capitan Land and Cattle Company, and had nothing to do with Lea. It did, of course, contribute to Thurber's losses in Lincoln County.

In the spring of 1889, the town of Roswell experienced a water shortage. The *Pecos Valley Register's* comment on the situation in its May 9 issue suggests how the local folks were indeed looking to Captain Lea as the 'Father of Roswell':

> The people of Roswell must be patient with the insufficient water supply at present. It is impossible for Capt. Lea to do the grading and work on the streets that is now going on and have ditches full of water. It will be but a short time until we will not be so troubled by a scarcity of water, when the work is done it will be a great improvement over the old system.

J.C. and Frank Lea and the Lea Cattle Company, together with Ada Edwards Davis and Richard F. Barnett became the owners in the 1890s of the old Woodlawn Ditch on the South Spring River. The priority for this water was established in 1872, even before John Chisum acquired the South Spring River Ranch, although the actual use of the ditch did not start until 1879, according to documentation at the Pecos Valley Artesian Conservancy District. John, James, and Pitser Chisum and Jim and Bill Miller originally started it. The Woodlawn Ditch was taken out of the north bank of the headspring of the South Spring River, about four miles southeast of Roswell. It was a large ditch, sixteen feet wide and two-and-a-half feet deep, and ran in a northerly direction for three to four miles. The 'V' scraper was undoubtedly used to enlarge and maintain the Woodlawn as well as the Lea Cattle Company ditch and other company ditches.

There were numerous other ditches besides those connected with Lea and his cattle company. Physical changes in the ditches were common as old routes were adjusted or abandoned and new ones excavated, new lateral ditches were dug, and constant maintenance was done on the ditches and related equipment. One of the most important changes resulted from the discovery of artesian water in Roswell in 1890, which provided a new source of water for the ditches. Shifts in ownership of the land and water rights provided another facet of unending change over the years. Certainly the development of irrigation ditch systems by Lea and others was a crucial element in the agricultural development of the Middle Pecos Valley.

A New Partner: Mabel Doss Day.

While he was in the midst of developing his ranch and related projects, Captain Lea's personal life took a new turn when he married for the third time. In 1885, Lea had gone to Texas to buy cattle when the Lea Cattle Company was forming. There he met Mabel Doss Day, widow of Colonel W.H. Day and proprietor of a large cattle ranch (77,550 acres) in southwestern Coleman County. Her grandson, James T. Padgitt, writes (*West Texas Historical Association Year Books,* 1958 and 1959) that she was known statewide as the 'Cattle Queen of Texas.' She was the only unmarried woman in the state who owned a big ranch—the largest ranch ever to be fenced in Coleman County.

Captain Lea pursued his courtship of Mabel Day for some four years, as did quite a number of rivals. He finally won her over, and they were married in her ranch house on April 29, 1889. Padgitt, who had the Day-Lea papers in his possession, states that Mabel turned around after the ceremony and spoke to the room full of cowhands: "I feel this is the greatest trade I have ever made and I believe I have made a good one. It beats cattle at $30 a head, although some of my old beaux don't think so!" With that, she kissed Lea, and the wedding party moved to another room to cut the cake.

Mabel Doss graduated from Hocker Female College of Lexington, Kentucky. She married W.H. Day in 1879 and they had a daughter, Willie Mabel Day, a year later. Unfortunately, Day died in 1881 without a will. The struggle that his widow endured to keep her property out of the hands of creditors is, according to

Gressley, "... unequaled for melodrama in the history of the cattle industry." Not only did she have to fight off the creditors, but she also had to deal with the avarice of her late husband's business associates and his brother, James M. Day. Many of those to whom her husband was indebted wanted to carve up her ranch and leave her out in the cold, instead of working with her to help her get out of a desperate situation. Finally, she got Professor Hocker and two Kentucky distillers to provide capital in exchange for title to all of her cattle. This provided the funds she needed to pay off her huge debts.

Then came the Fence-Cutting Wars. Some ranchers were opposed to fencing the range, especially with barbed wire. Anti-fence forces cut the fences that had been built on some ranches. Mrs. Day's ranch had extensive fencing, and Coleman County was right in the middle of the affected area. It took the Texas Legislature and the Texas Rangers to stop it, and Mrs. Day was left with many miles of destroyed fences when the cutting subsided in 1884. Her fence repairs, plus her personal debts, totaled $29,000—a huge sum in those days. She also owed thousands on the state land her husband had bought. A newspaper account said she came out of the scrape "... triumphantly, a free, independent and happy woman." However, she wrote to a friend, "I still owe better than $66,000 ... that's a long way from being free and independent .... I am bound to succeed. I will think of nothing else. I will go to the ranch and be so economical till I get out of debt." She spent most of the rest of her life fighting to keep her ranch and to expand her holdings, which she accomplished with considerable success.

Mabel Day Lea was still in the throes of debt when she married J.C. Lea in 1889. She turned the operation of her ranch over to her brother, William H. Doss; and she headed to Roswell with her daughter and her two black servants, Mattie and George. They rode the train to Pecos City, Texas. There, they loaded a wagon with the bedrolls and luggage. George drove the wagon, while Captain Lea drove a big hack carrying all the people. In Pecos, they met Gus Lee, a black hotel chef who had known Mabel Lea ever since she married Colonel Day in 1879. He joined the entourage and journeyed to Roswell to become their cook. He and Mattie later married.

When the party reached Eddy, they were welcomed by Charles B. Eddy and invited to spend the night with him. Eddy and his brother, John, were embarked on a development of their own in the lower Pecos Valley around the town of Eddy (changed to Carlsbad in 1899). Just before Captain Lea married Mrs. Day, Eddy had written a letter of encouragement to her, according to Padgitt.

> I am confident that your life here will be a happy one, for Captain Lea is one of the best men I ever knew. His character and reputation is excelled by none. He is very kind hearted, liberal and generous to a fault, extremely popular and has a host of friends. He is a large land owner in Lincoln County and stands as high as any man in the Territory among business men. I have known him many years and can say he is true and honorable in every sense of the word. And may I welcome you to New Mexico. Our valley is fast settling up with a pleasant class of people and I am sure you will be delighted with our climate.

The new Mrs. Lea's three black servants were downhearted when they saw Roswell, which consisted of a half-dozen adobe buildings and a street that resembled a cow trail. The Leas moved into the old hotel where the Captain had lived with his first family. According to Padgitt, their kitchen and dining room were in the back of the store building next door. Little Ella was brought back from California to reunite with her father and brother Wildy and to meet her new stepmother and stepsister.

Lea's ranching partner, Horace Thurber, highly approved of the Lea-Day marriage. On May 28, 1889, he wrote, "The news of your marriage has just come. I am delighted as I know a good wife if treated right, can be a wonderful helpmate. Please give my best to Mrs. Lea." Why his own wife lived in California while Thurber lived in New York is not clear.

Lea and his wife were absent from the home much of the time as they traveled, usually alone, in the interest of their separate holdings. Mabel Day Lea's experience and talents in ranching and management were invaluable in her new partnership with J.C. Lea.

Lea Cattle Company Hits Hard Times

Following Lea's marriage to Mabel Day, it was not long until he was again on the receiving end of Thurber's scalding letters. Thurber wrote on December 23, 1889 (cited by Gressley):

> I have your letter with bill of Mssrs. Poe, Lea & Cosgrove, and it just astonished me. My dear Captain, you keep telling me that you will stop drawing and stop these big expenses and yet they keep coming, and I have to pay so much money that I do not know where to get it; and it simply must stop, Captain.
>
> Mind you, Captain, I am not blaming you, but I am only trying to impress it upon you that when you are getting $7 or $8 a head for calves a year old, you can't have the same expenses as when you got high prices for cattle.
>
> I would like to know to what account to charge these groceries. I would be very glad not to complain, but I simply can't meet your demands, and your improvements have got to go; you have got to let them go by, and I have written you this so many times and I am almost ashamed to write it again.
>
> Now, Captain, to go back to one thing: you are paying the same prices you formerly paid for help and you are paying the same prices for groceries, now when cattle are down in price, that you did when they were high, and it would seem to me that some of you must come to grief (as I have said before), going on that kind of a scale; and either you have got to run your cattle at a lower rate of expense or else there is no profit in it; and it seems to me that all of you men will have to get together and simply put the price of wages down and have less average expense – and that is only a necessity.

Lea, tired of Thurber's harsh letters, entreated Thurber to buy him out as the only way to settle their differences. On December 23, 1889, Thurber wrote, "In regard to buying you out, Captain, I haven't the slightest wish to do anything of the kind. How could I get along with that great big property?" Gressley concludes that Thurber stubbornly refused to buy Lea out, "... on the ill-founded logic that putting up with a spendthrift Lea was better than having a headless operation or than attempting to train a new manager." Lea kept up the buy-out proposal in the 1890s.

One of the principals in the early days of the Lea Cattle Company, Charles S. Thurber, suffered an untimely death and was thus removed from the operation. Thurber died in July 1890 and was buried in Roswell.

There was a brief growth in the cattle industry from 1889 to 1891, recovering somewhat from the collapse in the cattle market of 1886-87. J.C. Lea, while shipping cattle in the fall of 1892, perceived that economic problems were ahead. He certainly was conscious of his own shaky finances. On September 17, 1892, he wrote to his wife from the Hotel Hagerman in Eddy (cited by Gressley):

> I see where Francis Smith & Co. are selling out land and cattle in Southern Texas and I am afraid of our Kentucky stockholders and Dallas ones... who would like our valuable property. The jealous hearted would be so pleased to see us in trouble. We ALL owe too much money, watch your expenses carefully.

Improvement in the finances of the cattle country was painfully slow. A severe drouth hit the southern plains in 1892-93, adding to the problems. Finally, the Panic of 1893 ended the recovery.

The Lea Cattle Company board of directors was already taking drastic action to escape financial ruin. The book of minutes shows that the cattle company entered into an agreement on October 22, 1892, with Charles B. Eddy, who was acting as agent for J.J. Hagerman of Colorado Springs. Hagerman would buy all of the company's land and the accompanying water rights in Chaves County outside of Roswell. The amount of land was estimated at thirteen thousand acres, and the selling price was $100,000. The transaction consisted of more than sixty-five separate tracts of land, plus five-sixteenths of the Pioneer Ditch, one-fourth of the Lea-Cunningham Ditch, two-thirds of the Hondo Falls Ditch, five-sixteenths of the Lea-Cockrell Ditch, one-fourth of the Roswell Town Ditch, one-and-a-half shares of the Woodlawn Ditch, and all of the South Spring River Ditch.

Minutes reveal that a special meeting of the stockholders was held at the law office of G.A. Richardson, "... the old office of said company having been previously abandoned," on Tuesday evening, January 24, 1893. The purpose of the meeting was to ratify the agreement with Hagerman to buy the company's lands and water rights. With the deal approved by the stockholders, the process of title clearance began. Finally, according to Chaves County deed records, on March 3, 1893, the purchase of 11,003.73 acres was completed; this was not "all" of the company's lands in Chaves County, as called for in the earlier motion. Other tracts were sold to Hagerman over the years: six tracts in 1894, two in 1895, and one each in 1898, 1899, and 1900. Also, about fifteen tracts were sold to other buyers in 1893.

The reason the company began to sell off land seems to be that Thurber at last gave up hope of ever recovering any of the hundreds of thousands of dollars he had invested in New Mexico, not to mention the twelve percent return he had wanted earlier. He refused to continue pouring good money after bad; Thurber stopped all funding and let the ranches fend for themselves. Indeed, the *Illustrated History of New Mexico* (1895), which is notorious for factual errors, avers that Lea served as president of the Lea Cattle Company "... until 1892, when he disposed of all his cattle interests."

Apparently, the company's Chaves County business after the 1893 transactions consisted of buying and selling farmland and city lots. Deed records show the company bought about two dozen tracts after 1892. Tracts sold in 1893-95 consisted of more than thirty-five transactions, plus another twenty-four over the next five years; thirty-five of the 1893-1900 transactions were for city lots. The last recorded sales were on February 14, 1900; and J. C. Lea as president of Lea Cattle Company signed all of them. Gressley states that the Lea Cattle Company went into liquidation in 1898, but there is no evidence to support that assertion. The company still owned thousands of acres of land in Lincoln County, but in the absence of extant records it is not clear what the status of the company ranch was in that county.

Court records show that with respect to the 150 bonds issued to H.K. Thurber on July 1, 1885, the cattle company did not redeem any more coupons, due semi-annually, after about 1891. Further, in a letter J.C. Lea (located in his probate records) wrote to his wife in November 1903, he indicates that he was disengaging from his business connections and planning to move to Texas where his wife was living. He expresses great anticipation of moving to Texas to be with Mabel and Willie:

> My dear Wife, I never could dare to promise myself any pleasure of this kind before because it was always so far in the distant future. But now I can see my way out & it makes me so happy to be able to indulge in the fond hope. Before, I did not dare to hope, only in the obscure distance. Now, my darling, it will not be long. I will work with all possible speed ... . It's one thing certain, however, that you can depend on my coming, & it will be before Xmas too, unless something happens or I get sick ... . I am so pleased at the idea that I can hardly hold myself. But don't say anything until I get there & I will surprise them all. I have been hoping for this [a] long time but did not want to tell you for fear something would happen & cause me to disappoint you & disappoint myself ... . I wonder where you are ... .

What these developments portended for the future of the Lea Cattle Company is unknown.

According to Gressley, the El Capitan Land and Cattle Company lasted until 1899. The only ranch that proved profitable for Thurber was the Arbuckle PO Ranch in Wyoming. Thurber earned the dubious distinction of being one of the biggest financial losers in the cattle-grazing business.

9

ORGANIZATION OF CHAVES COUNTY AND ROSWELL

Creation of Chaves County, 1889-1891

By the late 1880s, the development of the Pecos Valley from Roswell to Eddy brought about a demand for new counties. The huge size of Lincoln County made it very inefficient and inconvenient for citizens of the Pecos Valley to do business at Lincoln, the county seat. People around Roswell wanted a new county with Roswell as the county seat, and settlers in the vicinity of Eddy wanted a new county with Eddy as the capital. Captain J.C. Lea took the lead in the movement for the new counties.

According to George Curry in *George Curry 1861-1947: An Autobiography* (1958), dividing Lincoln County was the dominant issue in the legislative election of 1888. The Lincoln County Republican convention endorsed Frank Lesnet of Roswell for the Territorial House of Representatives. The district convention chose James J. Dolan of Lincoln for the Council (upper chamber of the Legislature); Dolan opposed the new counties. These two men were elected to represent Lincoln County. Contrary to what William A. Keleher states in *The Fabulous Frontier* (1945), J.C. Lea was not elected to the Legislature.

The county issue received much attention when the Legislature assembled in Santa Fe in January 1889. Captain Lea gathered a committee consisting of himself; former Sheriff Patrick F. Garrett; and the great developer of the lower Pecos Valley area, Charles B. Eddy. They traveled to Santa Fe to make the case for two new Pecos Valley counties. Democrat Governor Edmund G. Ross supported the idea, but the Republican Party controlled both houses of the Territorial Legislature. Lesnet and Dolan advised the men to go back and get Lincoln County citizens to sign a petition showing they supported the division of the county.

To carry out Lesnet and Dolan's instructions, Lea asked Lincoln County Clerk George Curry to call a public meeting. He described the boundaries of the proposed new counties to those assembled, and the people generally approved the plan and signed the petition. They appointed Curry to accompany Lea back to the Legislature to lobby for the new counties. Lea and Curry met with the governor, who agreed to sign any bill the Legislature passed. .

The February 17, 1889, issue of the *Pecos Valley Register* opined that the new county should be established by January 1890, which would "... stimulate business in every branch of trade." The paper also stated:

> Capt. J C. Lea has at last returned from Santa Fe, where he has been lobbying on the county division question. Much credit is due to the captain for the diligence with which he has followed this matter to a successful issue.
>
> He has spent lots of time and money, and we are glad to say will reap his reward as he should. There is plenty of work in the interests of Roswell to be done yet, and the captain is ready to take off his coat and wade in, and not let up until we have a city of recognized importance.

Two important changes were made in the bill while it was in committee. It went to committee as Bill No. 140, "An Act to Create the Counties of Roswell and Eddy," as shown on a document in the state archives. There was agreement that the southern county should be named for Charles B. Eddy, but there was no consensus on what to name the other county. Who chose to put the name 'Roswell County' in the bill is unknown. Many people lobbied to have it named 'Lea County' in honor of Captain Lea, but he resisted all effort to have it named for him. In committee, the name 'Roswell' was marked out by hand in more than twenty places where it appeared in the bill and the name 'Cháves' written in its place.

Lea insisted his county be named for Colonel José Francisco Chaves, a member of an old New Mexican Hispano family. Chaves was a former delegate to Congress and had served as president of the Council for several terms. Some believed Lea wanted to name the county after Chaves because this connection with a powerful Hispano would result in political advantages for the new county. Curry states that this step was taken to secure the votes of Hispanic members of the Legislature.

Will Robinson, editor of the *Roswell Register-Tribune*, in an undated article in the Lea papers in the archives of the Historical Center for Southeast New Mexico in Roswell, presented another take on how the county got its name:

> When Lincoln County received its major mutilation, the territory was overwhelmingly Republican. There were but a couple of Democrats in the Senate, and they had such an inferiority complex that they hesitated to start anything when asked to promote the creation of a new county with Roswell as the capital. Finally realizing that the approach was faulty, Captain Lea went to Col. J. Francisco Chaves, who held the Republican Party in the palm of his hand, and told him the story. He was courteously listened to, and when the story was told, Col. Chaves simply remarked that the request was right, and that he would see that the bill was passed. Captain Lea, in presenting the request thought it wise to say that it was the intention to name the child "Chaves County," in recognition of the distinguished service of the Republican leader, and incidentally, on the side, in appreciation of the help given. The answer came near being a classic.
>
> "Captain Lea, of course I would appreciate the honor you would confer on me, but it is not in my opinion either necessary or wise. I am a former Federal soldier. You served with the Confederates. In the nature of things, your new county will largely be settled with Southern people, of Confederate lineage. They could not be expected to be enthusiastic about having the name of a former enemy, and one of my race, used as the titled honoree. Such help as I have been able to give was rendered because your cause was just and not for any honor in return. It were better to select some other name."
>
> The answer was so fine that it appealed to the sense of chivalry of the old Confederate soldier, and he insisted successfully

Colonel José Francisco Chaves (1833-1904),
Chaves County's namesake.
Courtesy: HCSNM

> that it would be Chaves County, and nothing else, and it was so written. Col. Chaves retained a lively interest in Chaves County until the day of his death by an assassin's bullet [1904].

Curry records a story in his autobiography—and other writers have repeated it—that research shows to be untrue. He claims those who wanted to create Chaves County were mostly from Texas and did not want to use a Spanish name for their county, did not want to name it after a 'Mexican'—even one as prominent as Colonel Chaves. They compromised by spelling it with an 's' instead of a 'z,' which Curry erroneously states, is "... the correct spelling of Colonel Chávez's name." Historical sources reveal this story is false; this branch of the Chávez family has always spelled the name 'Chaves,' although they are related to those with the other spelling.

There has been other confusion about just exactly for whom the county was named. The problem stems from a story in the *Pecos Valley Register* of March 14, 1889, which mistakenly claims the county was named for Felipe Chaves of Belen. This Chaves was highly educated, very progressive, and very rich. The *Register* had already reported on February 19 that "The name is given in honor of Col. José Francisco Chaves, who we are glad to state was an honorable and virtuous man." How the newspaper afterwards decided that the county was named for Felipe Chaves is unknown, but others have perpetuated this erroneous assertion.

The other change that was made in committee concerned the county seat of Eddy County. The original bill designated Roswell and Eddy as the county seats of the respective counties. Citizens in the prospective Eddy County apparently wanted a choice about the county seat, so the bill was changed to designate Eddy "... unless a majority of the voters of the County of Eddy shall elect some other place than the town of Eddy for the county seat of said county" (Eddy carried the election.) The bill became law on February 25, 1889.

The western boundaries of the new counties were described in detail in the bill: "Commencing at the northern boundary line of the county of Lincoln on the line between ranges 19 and 20 east, thence south ... ." The line followed township, range, and sections lines with many zigzags, culminating on the "north boundary of the state of Texas" at the point where the line between Ranges twenty-one and twenty-two intersects the boundary. The county line falls approximately twenty-five miles west of Roswell. That part of Lincoln County west of the new boundary "... shall be and remain in the county of Lincoln."

The document further states that the third standard parallel south was the line between Chaves and Eddy counties, some thirty miles south of Roswell. Both counties extended eastward to the Texas state line initially, although new counties later resulted in the loss of much of the original territory of both counties.

Republicans in the Legislature had reservations about the usual practice of allowing the governor to appoint the officials of new counties who would serve until a regular election was held. They did not want a Democratic governor to appoint Democrats to the new county offices. This resulted in the unusual step of naming in the county creation bill three Republicans as temporary commissioners of Chaves County. The governor did appoint the temporary commission for Eddy County. Their duties were to see that elections were held, a courthouse and jail were erected, and other actions needed to prepare the new counties for launching by January 1, 1891. Those named in the law for Chaves County were W.S. Prager, W.H.H. Miller, and Henry Milne. The governor still had to issue their commissions, and he apparently was not quick enough. J.C. Lea wrote a letter on March 29, 1889, on Lea Cattle Company letterhead to Governor Edmund G. Ross (state archives) imploring him to act:

> We would be much gratified to have you to immediately commission the gentlemen named in the bill (the Act Creating Chaves County) as County Commissioners & forward them to this office.
>
> We are afraid to attempt anything more at present in the way of county organization ... . If we get our Commissioners at once we will have that much secure & will go ahead with the commencement of our buildings ... .

According to Curry, the Republican Legislature hoped the Republican commissioners of Chaves County would divide the commissioner districts and voting precincts in such a way as to favor Republican candidates. It did not work, however, as over the next few years many Democrats from Texas and the South settled in the Pecos Valley.

The block in downtown Roswell on the east side of Main Street between Fourth and Fifth Streets had long been referred to informally as the 'courthouse square,' because Captain Lea let it be known that a courthouse was what he hoped to see on that land eventually. In preparation for the forthcoming new county, Lea set out trees, sowed bluegrass, and put a fence around the square. Lea also hired the use of an Austin New Era Ditching and Grading machine from the Pecos Valley Irrigation and Investment Company in the summer of 1889, according to Dills. James R. Cunningham used a twenty-horse team to carry out the first highway grading in the Pecos Valley. The area graded was Main Street from the Río Hondo to the North Spring River. In June 1889, Lea made it official: he deeded the courthouse square to the new county. The commissioners accepted it and officially designated it as the site of the future courthouse.

The *Pecos Valley Register* of June 11, 1889, approved of the action and added its opinion about the building:

> Now comes the question of public buildings—what are they to be, brick or adobe?
>
> We suggest that of whatever material, they will be such as will be a credit to the county for years to come.
>
> It seems to us that it would be better to build something attractive and substantial and that may be enlarged or added to as the needs of the future may demand, rather than the attempt to spread the present appropriation [of] $30,000 over too much ground in buildings that will be considered shabby or useless in a few years.
>
> Let us have a good courthouse, gentlemen, something that we may look upon with pride after it is built.

The commissioners did not necessarily follow the newspaper's advice, as the building they erected was too small for enough

Chaves County Courthouse, built in 1890. This photo was taken in 1899.
Courtesy: HCSNM, Photo No. 747-A

growth and had to be replaced in twenty years. The contract for the building was let to J.P. Pauly & Sons of St. Louis, Missouri. They put up a beautiful building in a typical Victorian style. It was the first brick building in the Pecos Valley, according to the *Roswell Morning Dispatch* of September 17, 1948; and the bricks were made locally from local clay. The courthouse was two stories with a four-story tower in one corner. The jail was in a separate building behind the courthouse. The project came in at $29,000, which was a thousand below the amount appropriated in the creation law.

Lea wrote to the new governor, L. Bradford Prince, on November 15, 1889 (state archives), asking him to appoint Mark Howell as the government contract survey; "... I know him to be thoroughly competent." Howell, who later became a well-known educator in Roswell, was appointed. The elected commissioners later appointed him county surveyor.

The acting commissioners provided for regular elections to be held in the new county in November 1890. 'Judge' Edmund T. Stone, A. Brown Allen, and Walter P. Chisum were elected commissioners, with Stone as chairman. Voters chose Campbell C. Fountain as sheriff, Frank H. Lea as clerk, James X.Y. Sutherland as treasurer, and C.S. McCarty as assessor.

These men were sworn into office on January 1, 1891, in the new courthouse. Stone soon became quite ill, resigned on January 21, and died on the 25th. Allen took his place as chairman. The Board of Commissioners of the new Chaves County paid George Curry, Lincoln County Clerk, $1,144.34 to provide a copy of all county records pertaining to the new county. Further, the law required both Eddy and Chaves counties to assume their fair proportion of the indebtedness of Lincoln County.

Roswell Named County Seat, 1891

Between 1885 and 1891, Captain Lea was busy developing his ranch and getting Chaves County created. With the town of Roswell platted, he also sold town lots during those years and tried to attract productive people and businesses to his town. By the end of 1891, there were about a dozen businesses, three contractors, two physicians, and four lawyers in the town.

Roswell showed a population of 343 in the census of 1890; it was the logical choice for the Chaves County seat, there not being another 'town' in the county. Several developments added to the size and importance of the little village. A volunteer fire department was organized in 1889, and the Roswell Artesian Basin was discovered in town in 1890. This 'liquid gold' had an immediate beneficial effect on the town, the county, and the Pecos Valley.

Roswell leaders took steps to incorporate in 1891. Roswell's representative to the Territorial Legislature, attorney Granville A. Richardson, wrote a bill that the Legislature passed as "An Act for the Incorporation of Towns and Villages" on February 14, 1891. One source states that towns had to have a minimum population of five hundred. Perhaps this was a later requirement, because Roswell had only about four hundred in 1891.

Roswell residents immediately set forces in motion to comply with the new law. The first step was to prepare a petition signed by the majority of the qualified voters in the town. Such a petition was presented to the Chaves County Board of Commissioners at their meeting of June 1, 1891. When they determined Roswell met all the requirements of the law, they declared "... that said town of Roswell be, and is, an incorporated town ... ." The commissioners then proceeded to call an election for Monday, July 6, 1891 (James D. Shinkle, in *Fifty Years of Roswell History, 1867-1917*, erroneously gives the date as July 4th), at the courthouse for the purpose of electing a five-member 'Board of Town Trustees' and a town marshal. No registration was required, but only "bonafide residents and legal voters" (men twenty-one years of age or more) could vote.

The election was duly held, and the commissioners canvassed the votes on July 11. Town trustees elected were Nathan Jaffa, Smith Lea, Frank Lesnet, Steve Mendenhall, and Dr. Edwin Skipwith. Henry Wright was chosen as town marshal.

The town included not only the original townsite as platted in 1885, but also "South Roswell" and "other additions," according to the proceedings of the town trustees. Charles B. Eddy, president of the Pecos Valley Townsite Company, created South Roswell on March 10, 1891. The land, bought by Eddy from the Lea Cattle Company, extended from present Alameda Street ('Ash' at the time) seven blocks south to Summit Street, and from Kentucky Avenue eight blocks east to Sherman Avenue.

Actually, there were other additions to the town before South Roswell, all established by the Lea Cattle Company as parts of their land development and colonization program. They created 'Thurber's Addition' on October 25, 1888, extending from Second Street south to present Alameda and from Pennsylvania Avenue four blocks eastward to present Grand Avenue. Included in the plat was present Richardson Avenue, which at the time was called 'Overton'; and present Virginia Avenue, which was originally 'Pecos.'

They established the North Spring River Addition on May 23, 1889. It comprised the area north of the original townsite from Eighth Street to Fourteenth Street and from Pennsylvania Avenue four blocks east to Grand Avenue. That part of Grand Avenue became the route of the railroad in 1894 and 1899.

Another early addition was called 'Ovard's Addition,' which was filed on May 29, 1889, situated between Second Street on the north, Alameda Street on the south, Washington Avenue on the east, and Union Avenue on the west. Lea created the West Side Addition on May 27, 1890, for the cattle company. This was a large area extending from Fourteenth Street in the north fifteen blocks south to Alameda Street and from Pennsylvania Avenue four blocks west to Washington Avenue.

With these additions, then, the original town of Roswell as incorporated in 1891 encompassed essentially the area between Fourteenth Street in the north to Summit Street on the south, and from Grand Avenue on the east to Washington Avenue on the west, plus the nine-square-block Ovard's Addition which extended the town limits west to Union Avenue and south of Second Street. Of course, much of the area encompassed by the additions was vacant land at the time; but it represented the vision, hopes, and plans of J.C. Lea and many other people who believed in the future of Roswell.

At the Roswell Board of Town Trustees' organizational meeting on July 14, Nathan Jaffa was elected chairman. Records of the proceedings of the trustees indicate that they immediately went about the matter of running a town, setting up committees to deal with streets, sanitation, and other concerns of a town government. They were very diligent and precise—except for Mendenhall, who seldom attended a meeting—as they passed and amended several ordinances over the next few months. The first ordinance levied a five-mill tax on real estate within the town to provide funds for the functions of the town government. The second ordinance was for 'dog control.' Others ordered men to donate their labor to maintain the streets and required everyone on Main Street between Fifth and Walnut to put in sidewalks.

On September 23, 1891, Jaffa offered his resignation as chairman of the trustees, but it was not accepted until October 13. Frank Lesnet was chosen to take Jaffa's place. Around the first of 1892, Marshal Wright apparently moved out of town but did not resign from his position. The trustees fired Wright and hired C.B. Grant to fill the job on April 26, 1892. The town officials had a difficult time performing their duties during the last months of 1892, as none of their meetings from August 1892 until January 16, 1893, had a quorum present to conduct business.

With the county and the town both organized, the area was set for an influx of population, agriculture, and commerce. Captain Lea saw his dream of an economic boom in the Pecos Valley beginning to take shape. Although Lea was not much interested in holding office in the new county or town, the Town Board did appoint him on September 15, 1893, to attend a statehood convention in Albuquerque on September 20. It was to be many years, however, before statehood for New Mexico was finally accomplished.

10

ROSWELL'S DEVELOPMENTAL PERIOD

Because the story of Captain J.C. Lea is so intimately entwined with the development of the town of Roswell, it is necessary to include the story of the town in with Lea's story in order to place his activities in their proper context. Lea was always working behind the scenes to encourage those who were potential business developers for his town.

Jaffa, Prager & Company. The first large store in Roswell was Jaffa, Prager & Company, established in the fall of 1886 by Nathan Jaffa and Will Prager. Earlier that year they had opened a store at John Chisum's ranch, then moved it to town in the fall. The firm sold everything from groceries to clothing, so it was a true department store. It continued to grow for many years, with branches in other towns of Southeast New Mexico.

Churches. Helen Johnson started the first Pecos Valley Sunday School in August 1886. This led to the organization the following year of the first church, First Methodist Episcopal Church, South. They erected an adobe building at 311 North Pennsylvania Avenue in Roswell in 1888, which they used for about a decade. Several denominations began meeting for study and prayer, but it was several years before they established churches. Episcopalians started assembling together in 1890. The Disciples of Christ of Roswell organized First Christian Church in 1892, and the first minister was W.H. Bevill. The church built its first building at Fourth and Richardson in 1895. The Baptists and the Presbyterians also established churches in the 1890s.

Schools. The village school started on South Hill in 1885 became too crowded within four or five years because of a substantial growth in population in Roswell. In 1889 the Territorial Legislature enacted a law providing for public schools, so the town fathers built a larger school closer to the center of the community. They chose the block bordered by West Third and West Fourth Streets and by present North Kentucky and North Lea Avenues. Classes were held in the Methodist Church in 1890-91 while the new three-room brick (made locally) building was built. The 'Third Street School' was ready for the 1891-92 term; Samuel Hill and Alma Root were the first teachers in the new building.

Medicine. The first doctor in the Roswell area was Dr. James Sutherland, who homesteaded east of the Pecos River in 1882. Dr. Edwin H. Skipwith came to Roswell in 1887 from Las Vegas. He had to prepare his own prescriptions because there was no apothecary shop. According to Skipwith's relatives, he was born Edwin Henry Skipworth in Alabama; he changed his name to 'Skipwith' after he shot a man in Memphis and then decided it was prudent to go out west. Skipwith fabricated a background story to cover his past. Dr. William T. Joyner arrived in 1892, likely the best educated of the early doctors in Roswell. John Zimmerman opened the first drug store in late 1887 or early 1888. The Roswell Drug Store was near the northeast corner of Main and Second Streets. James Gilmore bought the store in 1890.

Newspapers. On November 29, 1888, J.A. Erwin and L.O. Fullen put out the first issue of the *Pecos Valley Register*, the first newspaper in Southeast New Mexico. James Kibbee changed the name to *Roswell Register* in 1890. Joe D. Lea (J.C.'s nephew) and Lucius Dills started the *Roswell Record* on March 6, 1891—the only Roswell paper to persevere until the present time.

Hotel Pauly. Hotel Pauly, built of adobe bricks, was a new hotel that opened in 1890 at present 114 North Main Street in Roswell. Some six months after the hotel opened, Joshua P. Church bought an interest in it and began to manage it. He and his wife Amelia lived in the hotel for about four years. The Pauly was considered to be a high-class place, the first 'real hotel' in the Pecos Valley. The office, dining room, kitchen, and verandah were on the ground floor; and there were eight bedrooms on the second floor.

In the hotel lobby, E.A. Cahoon and John W. Poe opened the Bank of Roswell in the summer of 1890. This first bank in Roswell erected an impressive building on the southwest corner of Third and Main Streets in 1894. In 1899, the name was changed to 'First National Bank.'

Stage Lines. Transportation in the Pecos Valley during the time J.C. Lea was developing Roswell was a problem. Most historical accounts, including maps, give little or no indication that stagecoach lines ever existed in the Valley. Some sources show that a mail route ran between Roswell and Colorado City, Texas, from 1870 to 1885. A regional line connected Roswell with Fort Sumner and the upper Pecos Valley starting in 1873, a shortline ran west from Roswell to Lincoln and White Oaks beginning in 1879, and a shortline started in 1880 that ran south out of Roswell down the Pecos toward Texas. Maps in the book *New Mexico in Maps* (1979) by Jerry L. Williams and Paul E. McAllister show these lines, but no other information about them has come to light.

An early stage line that connected Roswell with points west was Ozanne & Company Stage Line out of White Oaks, according to Robert L. Hart of the New Mexico Farm and Ranch Heritage Museum in Las Cruces. Urbain Ozanne received the contract in 1886 to carry the mail from Lincoln to San Antonio via White Oaks. The 'stages' were four- or six-passenger hacks pulled by two or four mules or horses. The line carried mail and passengers daily between Carthage and Lincoln. Connections could be made at Lincoln for 'Roswell, Seven Rivers, and Pecos Country' with the 'BUCK * BOARD * LINE.'

Pat Garrett, James Brent, and Steven Mendenhall developed a line of hacks and buggies as an adjunct to their land promotion and irrigation projects at Toyah and Pecos, Texas. In late 1889, they carried J. J. Hagerman on an inspection tour from Toyah to Roswell—a hack line between Pecos and Roswell was the result. Starting in January 1890, the line followed a schedule of three trips a week between Pecos and Roswell and back, with stops in Seven Rivers and Eddy. They carried mail and light freight, in addition to passengers. The 'stages' ran every day except

Sunday before the end of 1890 because of increased traffic (possibly only between Eddy and the railroad at Pecos).

Garrett and Brent sold their Eddy lines to Charles Eddy's town company in June 1890. The railroad reached Eddy from Pecos in January 1891 and eliminated the need for the hack line to Pecos.

The earliest information in Roswell about the stage line is in the January 15, 1892, *Roswell Record.* An advertisement proclaimed that Mendenhall, Barnett & Company Wagon Yard on the west side of Main Street had first-class accommodations for travelers. Business on the Roswell and Eddy stage line had been steadily increasing for the past month.

The 1892 *Record* also carried a timetable for the railroad between Eddy and Pecos and for the stage lines connecting Roswell with other towns. According to the schedule, a stagecoach left Roswell and another left Eddy at 6:00 a.m. every day except Monday, arriving at the other town at 6:00 p.m. the same day. Stations along the way were at Blake's Spring (later site of Artesia) and Seven Rivers.

Stages bound for Lincoln left Roswell at 7:00 a.m. every Tuesday, Thursday, and Saturday for the fifteen-hour trip. Stages returning from Lincoln to Roswell followed the same schedule except they traveled on Monday, Wednesday, and Friday. Stages on the 'Ft. Sumner and Roswell Stage Line' left each town at 6:00 a.m. on Monday, Wednesday, and Friday and arrived twenty-four hours later in the other town. La Mora Spring and Chaves Spring north of Roswell were stations along the route from Ft. Sumner.

Captain Lea's Social Activity

United Confederate Veterans. One interest that occupied Captain Lea's time was the Confederate veterans' organization. Roswell Confederate veterans were organized as the 'United Confederate Veterans, Valverde Camp No. 1419,' of which Lea was the Commander. The name came from the battle in 1862 on the Rio Grande in which Confederates routed Union troops at Valverde Crossing. Little has been learned about the organization; but it was considered to be a patriotic group that participated in public events and often heard orators on topics of patriotism, government, or politics. Similar camps were organized in other areas, such as the Bedford Forrest Camp No. 1606 in Portales. It is not known when the Valverde Camp was organized, but Captain Lea was the Commander as long as he lived. The organization continued to be active even after he died.

Masonic Lodge. One aspect of Captain J.C. Lea's life that has seldom been mentioned in historical accounts is his participation in the Masonic lodge in Roswell. The Roswell lodge was the first fraternal organization in the community and was a significant element in the development of the area. The first meeting to organize a lodge took place at the offices of Jaffa, Prager & Company on October 28, 1888. J.C. Lea was not among the four men who met, but his brother Frank was present. The others were Will Prager, John W. Poe, and William Fountain. They made plans to put up a building and to organize a lodge. Captain Lea donated a lot on the west side of the 200-block of North Main Street, and he and Frank each donated $50 toward the building. After much planning and work, Roswell Lodge No. 18 was granted a charter on January 20, 1890. Frank Lea served as the first secretary in 1889, Junior Steward in 1890, Junior Warden in 1891, and Senior Steward in 1893, according to the *Temple Book* published by the lodge in 1914. J. Smith Lea served in virtually all the offices at one time or another, including Worshipful Master in 1910. Other Lea relatives served as officers from time to time in the early years.

On the night of January 15, 1893, the lodge and everything in it burned. Fortunately, the secretary had taken the records home with him. Within days, the officers decided to sell the original lot that Captain Lea had donated and to build a new facility at the northwest corner of Third and Main Streets. This time, Captain Lea donated $100 to the building fund, as did Frank and Smith Lea. Apparently, J.C. Lea never held an office in the Masonic lodge, but he was in sympathy with its goals and supportive of its causes.

Church Membership. Although the records of First Christian Church from the earliest times have been lost, it is known that Captain Lea was a member of that church. No evidence that his wife or children were also members has surfaced, but it would be reasonable to assume that they were.

Wildy Brothers

Ernest L. Wildy. Ernest L. Wildy lived in Roswell in the 1890s and worked as a real estate agent for several years. The community accepted him as a leader and elected him to the Town Board of Trustees. He served a year as chairman of that board from April 1898 till April 1899.

Although records are sketchy, it is known that Wildy had a business partner named 'Smith' for a time around 1902 and was well established in the business. That same year he took on a famous partner: 'Sockless' Jerry Simpson, the noted former Populist congressman who had represented Kansas in the House of Representatives for three terms in the 1890s. Simpson moved to Roswell in 1902, where in July he and Wildy entered into a partnership. Their advertisement in the *Roswell Record* was as follows:

JERRY SIMPSON AND E.L. WILDY
DEALERS IN
REAL ESTATE AND LIVE STOCK
Irrigated Farms, Ranches, and Town Property for Sale

A later advertisement listed nineteen places to sell in Roswell; it gave Simpson & Wildy's office location as "upstairs, opposite Citizen's Bank." This location was on the southwest corner of Fourth and Main Streets. Simpson and Wildy dissolved their partnership about March 1903. Wildy next formed the 'Wildy Realty Co.' with partners C.J. Bowman and C.E. Edgerton. On July 10, 1905, Wildy was living in the Shelby Hotel when he became one of eleven local people to fall victim to typhoid fever. Wildy survived and may have gone to California soon afterward.

John H. Wildy. Sallie Lea's other brother, John Hall Wildy, also settled in New Mexico but lived mostly away from Roswell. A deed record in the Chaves County courthouse shows that he lived in Lincoln County in 1883. According to Joseph R. Anderson,

Ernest L. Wildy, Sallie Lea's brother, was a prominent real estate agent in Roswell in the 1890s and early 1900s.
Courtesy: Barbara Buckner, Carlsbad, New Mexico

who researched Virginia Military Institute (VMI) alumnae (probably in the 1920's), Dr. C. Wildy Bonney of Satartia, Mississippi, told him that John Wildy entered VMI in the fall of 1870:

> He was graduated at the University of Mississippi after leaving the VMI and became a professor of civil engineering at Los Angeles, Calif., and Carson City, Nev., and is said to have stood at the top of his profession and was in the employ of Western Railroads the greater part of his time. That he later had a cattle ranch at St. Augustine, near Las Cruces, New Mexico. Dr. Bonney says that Wildy married a Miss Davis .... He says he [John H. Wildy] died from being thrown from his horse at Miami, Texas, November 5th, 1893 ... .

William A. Keleher's account in *The Fabulous Frontier* (1982) avers that Wildy started a ranch in the Tularosa Basin south of present Alamogordo and had his headquarters at his well. Sources vary the name from 'Wildey Wells' to 'Wildy's Well,' 'Wilde Well, and 'Wildy Well.' Wildy's Well, Keleher avers, became more noted than either he or his ranch, because it was the place "... where the famous gunfight took place in 1898 between Oliver Lee and Jim Gilliland on one side and Pat Garrett and his deputies on the other."

Perhaps a more reliable source is an article in *Frontier Times* magazine by Marshall Hail in the August-September 1969 issue, "San Augustine Ranch House."

This house was at the eastern foot of the Organ Mountains, just south of the White Sands. One Thomas Bull established himself at San Augustine Spring just after the Mexican War, Hail states, where he prospered in lumbering and in buying and selling horses and cattle. He sold the place to Warren T. Shedd of St. Louis.

Shedd built a commissary and an inn; he also sold fuel and feed. Business was so good that he put in a bigger hotel, a stage station, and big corrals. He built a rock fence around the spring and started selling the water to travelers.

Gold was discovered at Brice, west of Orogrande on land that later was acquired by the San Augustine Ranch. The gold brought

Above: Early artesian well near Roswell. Man on right in derby is Captain J.C. Lea. *Courtesy: HCSNM, Photo No. 1888*
Below: Frank Lea's artesian well in Roswell flowed 300 gallons per minute. Second from left: Pearl Lea; fourth, Wildy Lea; fifth, Jennie Lea. *Courtesy: HCSNM, Photo No. 647*

new customers to Shedd's place, some of whom were lawless men who often engaged in gunplay. But Shedd enjoyed the income and built most of the Spanish-frontier-style main ranch house—fifteen rooms around three sides of a patio. The elaborate compound was also a fortress against Apache attacks.

After the Civil War, Shedd added a big dance hall, gambling tables, and a brothel. Shedd's ranch developed a reputation as a rough place where 'anything goes.' The outlaw gangs who roamed southern New Mexico, especially Lincoln County, in the 1870s, found the welcome mat out at Shedd's San Augustine Ranch. One such incident involved the theft of some of Richard Brewer's horses on the Rio Ruidoso by Jessie Evans' gang. Brewer's friends found the horses at Shedd's and recovered them.

One of Shedd's neighbors was sheepman Benjamin E. Davies, who wanted for many years to acquire the San Augustine. Finally, Davies made a deal with Shedd sometime in the late 1870s to purchase the ranch. Davies, from Wales, had come into the Mesilla Valley with the California Column in 1862 to fight the Confederates who were occupying the territory.

Davies and his wife, Julia, being refined and sophisticated people, greatly changed the nature of the ranch. The outstanding variety of debauchery offered by Shedd was brought to a screeching halt. Some of the most prominent people of southern New Mexico Territory were properly entertained at the Davies place.

Davies was a prosperous livestock producer, with large herds of both cattle and sheep. He died sometime in the late 1880s. This is where John H. Wildy enters the picture. Wildy secured the position of administrator and manager of the ranch. Mrs. Davies had a daughter, Jessie, by a previous marriage, who later married Wildy.

According to Hail, Wildy drilled his famous well eight miles east of Orogrande. Eugene Manlove Rhodes places the well at the southwestern corner of the White Sands in his novel, *Paso por Aqui;* but that is undoubtedly inaccurate. A U.S. Geological Survey map of the area made in 1943 shows 'Wilde Well' and 'Wilde Tank' some six miles east and five miles north of Orogrande; *i.e.,* about eight miles northeast, which is in Section 26, Township

21 South, Range 8 East. The well is shown just east of the 106th meridian and about four miles south of the 32°30' parallel. The website of White Sands Missile Range on the Internet essentially agrees with the Hail article, but it spells the name incorrectly: 'John H. Wilde.'

Some local sources do not support the claim that Wildy dug the well. Pete Eckermann of El Paso, an authority on the geology and geography of Otero County, maintains that German immigrants named 'Wilde,' which is the spelling used on official maps, probably dug the well. Eckermann also points out that the exact location of the well may have been obscured by Civilian Conservation Corps work there in the 1930s.

At any rate, Wildy's operation of the San Augustine Ranch lasted for just a few years because a relentless dry spell caused a great loss of cattle. Wildy took the rest of the cattle to the Texas Panhandle, where he met an untimely death. An article in the *Eddy Argus* on November 24, 1893, a reprint of an item in a Roswell newspaper, reported: "Last Sunday the sad news of the death of John H. Wildy reached his brother and sister here [Roswell]. He was killed at his ranch in the panhandle, by a horse falling with him." The brother and sister were Ernest L. Wildy and Ada Wildy Davis.

Jessie Wildy and her mother were unable to operate the San Augustine Ranch successfully, so they sold it to William W. Cox and it has been in the Cox family ever since. Cox and his son, James W. Cox, played important roles in the business, ranching, and political developments of the Doña Ana County area.

Goss Military Institute instructors and students, *c.* 1892-94. Colonel Robert C. Goss reclining on ground, center front by the artesian fountain. Taken at the original location of the school.

Photo by Phillips & Sheek. Courtesy: HCSNM, Photo No. 2120-A

11

FOUNDING NEW MEXICO MILITARY INSTITUTE

Roswell, the city whose early development was promoted largely by Captain Joseph C. Lea, is home to one of the nation's most outstanding military schools: New Mexico Military Institute. Lea was called the 'Father of the Institute' as well as the 'Father of Roswell.' The little military academy that J.C. and Mabel Lea coaxed into existence has for many decades enjoyed a reputation as an excellent school, both academically and militarily. Its continued success is a living tribute to Captain Lea.

When J.C. Lea married Mabel Doss Day in 1889, his son Wildy was twelve, his daughter Ella was eight, and Willie Day was nine. Willie was happy to have two stepsiblings, because she had not had many playmates living on the isolated Day Ranch in Texas.

In Roswell, the three children of the Leas attended the community school on South Hill, which was fast becoming over crowded. The South Hill school and the Farms School were the only schools in eastern New Mexico, and few public schools existed anywhere in the Territory. From the time they married, J.C. and Mabel Lea mulled over the problem of schooling for their children. One factor that no doubt sparked Lea's interest in schools was that on March 24, 1890, Governor L. Bradford Prince appointed him to a represent New Mexico in Washington to lobby the federal government to grant lands in the Territory to be used for public schools and colleges, irrigation reservoirs, and a national park in the Sangre de Cristo Mountains. The document of appointment is in the Lea papers. Whether he went to Washington or not is unknown.

Perhaps the military school at Roswell owes its beginning to the youthful rebellion of Wildy Lea. After five years without a

Ella Lea and Wildy Lea, children of J.C. and Sallie Wildy Lea; with Willie Day, their step-sister, daughter of Mabel Day Lea. Wildy is wearing the uniform of Fort Worth University. *c*. 1890. Photo by Swartz of Fort Worth.
Courtesy: HCSNM, Photo No. 495-C

mother, Wildy had acquired some undesirable characteristics from the rough cowhands with whom he associated; and his father became concerned. When Mrs. Lea was in Texas attending to her ranching business in March 1890, Lea wrote her a letter (cited by her grandson, James T. Padgitt, in "Early History of New Mexico Military Institute," *West Texas Historical Association Year Book,* (1958) that pretty much explains the Captain's problems with his only son:

> I am determined to put Wildy somewhere to be trained. I shall put him first in a convent until he is subdued and then later in a military school. I think you and I fully agree that he has to be placed entirely under strangers and out of the cowboy atmosphere. Meet me at the Pickwick Hotel in Fort Worth about April 1st.

Two weeks later, Wildy Lea found himself enrolled in the Junior Preparatory Department of Fort Worth University, where the Commandant of Cadets, Colonel Robert S. Goss, exercised resolute authority over his cadets. Over the next few months, Mrs. Lea made frequent visits to see her stepson as she came and went to her Coleman County ranch on business trips. She got to know Colonel Goss rather well. She was impressed with him and saw he was quite successful in 'subduing' young Wildy Lea.

While most accounts credit Captain Lea with being the moving force behind the establishment of a military school in Roswell, Padgitt indicates that she was at least as active as the captain, maybe more so. Padgitt had J. C. and Mabel Lea's letters and other papers, and those documents are still in the possession of the Padgitt family in San Antonio, Texas.

During the fall of 1890, when thirteen-year-old Wildy Lea was enrolled for a second term at Fort Worth University, Mrs. Lea proposed to Goss that he should establish a military school in Roswell. He did not think much about it at first; but when it was time for Wildy to go home for the Christmas holidays, Mrs. Lea and the Captain both arrived at the school to invite Goss to spend the recess with them. A bachelor with no particular plans for the holidays, Goss agreed to travel to Roswell with the Lea family.

Captain Lea at first envisioned a 'university' in Roswell that would not only have local importance but would affect the entire Territory, even the nation. He wrote to his ranching partner, Horace K. Thurber, to ask him to donate Lea Cattle Company land to put up such a school, in exchange for which the school would be named 'Thurber University.' Thurber replied as follows:

> In regard to donating the sixteen acres for the university, I approve of it if you think it the best thing to do and I leave it up to you. I do not want to have that called the "Thurber University." I think it would be a great mistake. It should be "The Roswell University of Chaves County." Then the title explains itself; it advertises the town.

Thurber later contributed money to buy books for the library of Lea's school.

Roswell's outlook was very promising at the end of 1890. The new Chaves County was about to be organized, and a new courthouse was finished and ready for the new government to occupy. Roswell, with a population of 343 in 1890, was designated as the county seat; and plans were underway to incorporate the town the following year. Artesian water was discovered while Goss was in Roswell, and that set off a boom of development and population growth. A new, three-room, brick schoolhouse would soon be built near downtown Roswell.

Goss appeared in a public recital at the South Hill school while he was in town. He rendered several well-known readings to a sizeable audience. He did not pass up the opportunity to publicly praise J.C. and Mabel Lea for their interest in promoting education in Southeast New Mexico. The Leas were pleased that their vision was moving forward.

Despite the favorable response from Goss, the whole idea of starting a military academy in Roswell—which claimed to be farther from a railroad than any other town in the country—seemed to be imprudent. Goss lost some of his enthusiasm when he returned to Fort Worth, but Captain Lea wrote numerous letters to him and Mrs. Lea visited with him frequently. Their efforts brought Goss around. When the 1890-91 term ended in June, Goss resigned from Fort Worth University and headed to Roswell to face the challenge of establishing a brand new military school.

A temporary site for the school was provided in the spring of 1891 when the Leas offered five acres of land and an adobe house on their property on the north half of the present 400-block of North Main Street, just north of the Lea residence and across the street west of the county courthouse. The Roswell community held a fund-raiser on June 22, 1891, in which thirty-six local citizens pledged $1,330 to build a study hall, a dormitory, and a six-foot-high plank fence around the campus. The contributors understood that the buildings would later be moved to a permanent site for the school. Before the school term started, local patrons organized a minstrel show and a concert to raise additional funds.

Some thought the board fence was an unnecessary expense, because times were hard and lumber was expensive. As Captain Lea's son-in-law, Hiram M. Dow, said in 1925 when he presented an account of the early history of the military school to the Roswell Woman's Club (newspaper article, n.d., *c.* 1925; Archives, Historical Center for Southeast New Mexico, Roswell), "We must remember that in that day and time, things just had be 'corralled.'"

Goss wanted plans for a facility that could accommodate fifty boarding students, and a board of advisors was organized to help him. The board included Captain Lea, businessmen Nathan Jaffa and Will Prager, bankers John W. Poe and E.A. Cahoon, and attorney Granville A. Richardson. Dow states that former sheriff Pat F. Garrett was also one of the advisors; he was not on the original board but served briefly later, according to William E. Gibbs & Eugene T. Jackman in *New Mexico Military Institute: A Centennial History* (1991).

Goss originally placed the name 'Goss Military Academy' on his letterhead, and some early newspaper items refer to it by that name. At least one document in the state archives uses the word 'Academy,' according to Ronald L. Peters in his 2001 Master's thesis about the campus development and architecture. Peters believes the name was changed to 'Institute' when the Territory decided to assume responsibility for the school. However, most early sources called it 'Goss Military Institute' (GMI) earlier than the time posited by Peters.

Gibbs & Jackman identify the earliest document in the archives of New Mexico Military Institute that pertains to the creation of GMI as a letter written by Captain Lea to former Roswell resident Frank G. Tracy, who had moved to Eddy. On August 14, 1891, Lea wrote:

> We are getting along fine with our military school building. But we are about to get out of money with which to finish it & I am requested by the committee to ask you to please send to Mr. [W. S.] Cobean your subscription if you have not already done so.

Goss successfully advertised the new school throughout New Mexico and parts of Texas. His prospectus stated bluntly, "If you want your boy managed, send him to Goss; if not, keep him at home." Goss, of course, was the superintendent. He later was given the rank of 'Colonel' and aide-de-camp as an assistant to the governor. Goss engaged a well-qualified faculty of five other men with impressive credentials.

According to J.R. Kelly in *A History of New Mexico Military Institute, 1891-1941* (1953), the first GMI facilities at Fifth and Main Streets were adequate but unimpressive. An adobe house belonging to Lea was near the campus entrance; and it served as a reception area, infirmary, and Colonel Goss' living quarters. Behind it was the main facility, which housed the classrooms and dormitory. It was a long, unpainted wooden building situated on an east-west axis. Another old adobe house was used for a kitchen and mess hall; the chef was Gus Lee, the Lea family's cook, assisted by his wife, Mattie, who was also the laundress. Behind the dormitory was a 'natatorium,' a fourteen-foot-by-twenty-five-foot edifice built across the J.C. Lea Ditch, a flowing irrigation *acequia* that no doubt was 'descended' from the original ditch on the Roswell townsite. Inside the building was a swimming pool

Female students of NMMI, February 1, 1894. r. to l.: Carrie Wilson, Willie Day, Ella Lea, May Neatherlin, Nellie Johnson, Jennie Lea, Ola Fountain, Winnie Neatherlin, Burnice Fredricks, Nora Wiggins, Eliza Fountain, Mabel Bull, ? Pollock. Photo taken in the original location of Goss Military Institute and NMMI, 400 Block of North Main Street at Fifth, West Side.
Courtesy: HCSNM, Photo No. 3344.

that Wildy Lea said was chin deep. GMI took pride that only one other school in the country had a natatorium, and it was smaller than GMI's.

The campus of the school featured a lawn dotted with cottonwood trees and a round artesian fountain. On the west side was a large area for military drills and parades. A small orchard encompassed a zoölogical garden, which was home to a lobo wolf and a black eagle. This first location of the military school was used for three years (despite Dow's assertion that it was used only one year).

This description of the campus raises a question about exactly where it was located and whether or not it extended north of Fifth Street. No source mentions an adobe house between J.C. Lea's residence and Fifth Street, although that does not preclude the existence of such a structure. It is known that the adobe house where William H. Cosgrove had lived in the mid-1880's, where J.C. and Wildy Lea lived with Milo and Ella Pierce during the period from 1884 to 1889, and probably where Frank Lesnet was living in 1891, was on the northwest corner of Main and Fifth Streets. Also, the earliest detailed map shows that in 1900 the J.C. Lea Ditch flowed along the north side of Fifth Street. It seems probable that the campus spread across and closed off Fifth.

On September 3, 1891, Goss Military Institute opened its first term with students from Roswell and all around New Mexico Territory. Just how many students there were is a matter of disagreement among the sources. Kelly writes that there were thirty-eight students enrolled, and twenty of them were girls. That would leave only eighteen boys. An early faculty member recalled no more than ten girls, according to Kelly. Kelly also states that the female department was separate from the rest of the school, which Padgitt takes to mean that there were thirty-eight boys plus twenty girls. The boys, nearly all of whom boarded at the school, paid $300 for the year, while the girls, local day students, were not required to engage in military activities to the same extent as were the boys, paid $90. The cadets' uniforms were West Point gray.

Gibbs & Jackman conclude that the reason girls were allowed into the military school was that Mabel Lea insisted that girls be included. Her own daughter, Willie Day, was ten years old; and her stepdaughter, Ella Lea, was nine by the time the school opened. They were among the students in the female department. The age range in the first student body went from seven years to the twenties. Having girls in the school necessitated the presence of a matron, and Goss hired Miss A.E. Hassen as supervisor of the female and preparatory departments. According to Kelly, the demand for "accommodations in the female department" made it necessary to let a contract for a 'Daughter's Home' within days after the first term started. However, Gibbs & Jackman maintain that girls never actually lived on campus in the early years.

Financial problems arose promptly in the first session of GMI. Colonel Goss wrote to Captain Lea on October 3, 1891 (NMMI archives, cited by Gibbs & Jackman):

> The school is doing fine, but no money. The people cannot pay, it seems, but we are living. If I can in any way hold up this year, I will be all right next. But I will get through somehow.

Two weeks later Goss wrote to Mabel Lea (NMMI archives, cited by Gibbs & Jackman) and urgently asked her for a thousand dollars. He tried to sound very upbeat in his letter:

> I pause to tell you what I hear in the next room—some fifteen small girls in voice culture. If you could hear the sound that rings in the air, you would not—could not think that this is Roswell, New Mexico. Oh! How I am fighting for the good work. If my health will only last me to the end.
>
> I am thinking of leaving for a trip of some four weeks to Lincoln, White Oaks, Fort Stanton and the ranches. I must get out and secure some ten more boarders for January 1st. I must have that much money to run me through.

According to Kelly, Goss Military Institute was incorporated in January 1892 by Goss, Lea, and five other men; but documents on file at the Chaves County Courthouse show that GMI was incorporated on September 12, 1891, for a period of fifty years. The papers were filed with the Secretary of the Territory on

October 3 of that year. Captain Lea provisionally donated forty acres of land on the banks of North Spring River, a mile and a half northwest of town, for a permanent campus (where the west end of Spring River Municipal Golf Course is presently located). On Arbor Day, the GMI students set out 250 shade trees on the proposed new campus. Obviously, the school never moved to the Spring River location, although the reason it did not is unclear. It seems that a restructuring of the school resulted in developments taking a different direction with different people in charge.

Another concert benefited GMI in May 1892. Among the performers were Mabel Lea, her daughter Willie Day, and Captain Lea's niece, Jennie Lea. Jennie, as well as Ella and Wildy, were often listed among those appearing on various programs presented by the school.

During the first year that GMI was in operation, there was gossip around Roswell that Miss Hassen was romantically involved with Colonel Goss, according to Gibbs & Jackman. Mabel Lea, shocked at this "unbelievable" rumor, insisted that Goss terminate Miss Hassen's employment. Goss declined to dismiss her because he felt that that would be an admission of guilt. Dr. Gibbs, in a reference note, allows, "Mrs. Lea's work in behalf of education and civic betterment was well known, as was her inclination to delve into other people's affairs" (no pun intended). Unfortunately, Miss Hassen died during GMI's second year.

The 1892-93 school year saw some sixty students enrolled at GMI in eight grades of schoolwork. The original music teacher resigned, so Colonel Goss hired Julia Fitch of Albuquerque to teach music and to serve as matron. The superintendent and Miss Fitch were attracted to each other and soon were making plans to marry. They planned a wedding in June 1893 at her home; however, she soon found that "... she could never be matron and escape the town's love of gossip." The Reverend J.D. Bush married them on December 9, 1892.

At the same function in which Goss and Miss Fitch were married, the first cadet officers were installed. Among the several officers were James M. Hervey, future district attorney, as second lieutenant; Wildy Lea as third sergeant; and Edgar Calfee, Captain Lea's nephew, as assistant drum major.

In a letter to E.L. Lusk in 1951 (NMMI archives, cited by Gibbs & Jackman), Mrs. Goss stated that Goss' refusal to fire Miss Hassen led to a break between the superintendent and his main backer, Mabel Lea. Mrs. Lea did help organize programs of entertainment, the Leas even boarded at the school (which was next door to their home) for a part of the school term, and young Wildy Lea continued as a GMI cadet; but the previous close relationship between Goss and the Leas ended.

Goss' military academy had been on shaky financial ground from the beginning. Gibbs & Jackman conclude that the tensions between Goss and the Leas "... probably led to the rescue of the foundering school" because the perceived 'drift' of the school was a signal to the people in the Roswell community to take steps to shore up the school, both financially and institutionally. Dills wrote a brief history of Roswell in 1932 in which he credits Captain J.C. Lea (who was Dills' wife's uncle) with a "... vision that was truly prophetic" in seeking to establish a "Military Training School" that would become "... statewide, if not national in its scope and influence." By contrast, Dills writes, from Goss' viewpoint, "... the school was a personal affair, with himself as the personality."

In early 1892, Captain Lea contacted Antonio Joseph in Washington, who was New Mexico's Territorial Delegate to Congress, to ask him to seek federal assistance for the fledgling military school. Joseph wrote to Lea (cited by Padgitt) on March 15 to report his failure:

> I made a strong fight before the Committee on the Territories for the Goss Military Institute, but lost it. The Committee opposed all grants of public lands, except to state institutions. If Goss Military Institute was now under the control of the Territory, there would not be much trouble to secure a grant of land for it, but under existing circumstances it cannot be secured.

Joseph's letter settled the issue in Lea's mind: GMI was "dead beyond resurrection," as he later put it in a letter to his wife. The school had to be recognized by the Territorial government in order to obtain financial support from any level of government.

Accordingly, Captain Lea asked Dills to draft a bill to submit to the 1893 Territorial Legislature providing for the "absorption of the Goss Institute and the creation of the 'New Mexico Military Institute.'" Some sources give credit for drafting the bill to E.A. Cahoon, or to Dills and Cahoon together. In any case, Lea was behind the action. Goss was not involved; he claimed to be surprised when he learned of it. Dow relates a story that had been passed down in the Lea family that Captain Lea drove in his buggy to Lincoln, where he picked up out-going Lincoln County Sheriff George Curry (a future governor). The two of them went to Santa Fe to lobby for the passage of the New Mexico Military Institute bill. Actually, the original bill would have made Goss Military Institute a Territorial institution without changing the name; but A.B. Fall suggested calling it 'New Mexico Military Institute' in order to garner support from other areas of the Territory. This was done, and the bill was approved on February 23, 1893; it made no provision for financial support for the school.

Governor L. Bradford Prince immediately appointed the first board of regents for New Mexico Military Institute: Goss, Lea, John W. Poe, Nathan Jaffa, and Charles B. Eddy. On July 14, 1893, Governor W. T. Thornton, who had replaced Prince, appointed Lea "Aid[e]-de-Camp" with rank of Major, Volunteer Militia of the Territory of New Mexico, according to a document in the Lea papers. It is believed that the governor likely appointed all of the NMMI regents as Aide-de-Camp. Governor Thornton attended the first meeting of the regents in Roswell on October 8, 1893, in which Jaffa was chosen chairman. Lea was reappointed to the board periodically for the rest of his life.

Gibbs & Jackman write:

> Despite the fact that the nature and the name of the school had been changed without the foreknowledge of its superintendent, there is no evidence that those involved necessarily intended to get rid of Goss. In fact there is evidence, despite

> Mrs. Lea's fantasies about his social life, that he might have stayed as long as he desired ... .

Goss did not elect to stay, however. He announced his resignation before the school term ended in the spring of 1893. While he expressed no resentment at the time about what the Legislature did, he later hinted that he felt he did not receive appropriate recognition for his part in founding the Institute.

The regents hired J.E. Edgington, who had joined the faculty in 1892 as an instructor of classical languages, to succeed Goss as superintendent. Dow takes the position that the school continued for two more years as a private institution without government aid. The main difference was the change in name. The regents met for the first time in October 1893 and apparently did not meet again until April 1895. The regents told Edgington to keep the school in operation to the best of his ability and to take whatever funds he could earn from it. Like Goss, he was continually in financial distress.

The first school year with the new name and under Territorial sponsorship, 1893-94, was conducted in the original location at Fifth and Main Streets. Fifty-four students enrolled for that term, according to Gibbs & Jackman; and twenty of them were girls.

In early August 1894, the original Goss Military Institute frame buildings were moved from the 400-block of North Main Street three blocks north to the west side of the 700-block. Presumably, it was possible to reconstitute the natatorium in the new location, because another ditch, the Pierce, Cunningham, & Ballard (Centre) Ditch, crossed the townsite just north of the new campus at Eighth Street. However, none of the accounts mention the natatorium in the new location. Kelly maintains that an unexpected rise in the number of girls enrolled created a need for more housing, so in November construction was started on a large dormitory and parlor built of adobe bricks just south of Eighth Street. Again, Gibbs and Jackman maintain that Kelly is mistaken and there is no evidence that girls ever actually boarded at the school during the early years.

The records of enrollment for 1894-95 are lost, but the school year began on schedule on September 12. A month later the

Wildy Lea, wearing NMMI uniform. Photo by Frank Frost.
Courtesy: HCSNM, Photo No. 3341-C

NMMI with first Lea Hall Headquarters and four other buildings, *c.* 1900.
Courtesy: HCSNM, Photo No. 165-U

cadets paraded in town as part of the celebration of the arrival of the Pecos Valley Railway in Roswell.

The day-to-day operation of the school was a constant effort to say afloat financially. It appeared to be only a matter of time until NMMI would have to cease operations, although steps were taken to rescue the school. George Curry, who was then a member of the Territorial Council, met with Captain Lea and other supporters of NMMI in late 1894. They drafted Council Bill No. 40; Curry introduced it in the Legislature and Representative James F. Hinkle helped carry it to passage. It was passed on February 13, 1895, and authorized the Institute to issue bonds for buildings and improvements. It also included NMMI among the territorial schools to be partially supported by a tax levy. According to the *Roswell Record* of March 1, Curry telegraphed Lea: "Legislature adjourned. Roswell gets $7,000 for Military School."

These measures were not completed in time to stave off disaster. The end finally came on March 29, 1895—a day known in NMMI history as 'Black Friday.' The school was forced to stop operations, and it did not resume classes for three years. In the summer of 1895, one of the NMMI buildings was converted into a laundry and another became a boarding house, Kelly states.

Although the Institute may have appeared to be dead, it certainly was not dead in the minds of local and territorial supporters. Much activity continued during the time the school was closed. Another provision of the bill passed in February 1895 required the school to secure the donation of at least forty acres of land for a permanent campus. Gibbs & Jackman state that no explanation was forthcoming for why the regents did not accept Captain Lea's offer of forty acres on the North Spring River. Instead, they sought a donation from developer J.J. Hagerman of land located just west of present North Main Street at College Avenue on North Hill ('The Hill'). Other lands were added later from time to time.

There was about a year delay in selling bonds because of the need to secure the approval of Congress and the president. Even before the bonds were sold, the regents began developing the site and securing a water system. The bonds were sold in June 1897, and that same month a contract was awarded for the erection of a three-story building to cost about $12,000. In December 1897, according to Gibbs & Jackman, Captain Lea requested that the regents promptly choose a superintendent from those applying and let the other applicants know their choice. Early in 1898, the regents hired James M. Meadors for the superintendent's position, with instructions for him to be on the job by March.

The regents decided in April 1898 to allow some female day students to enroll, Gibbs & Jackman state. In May, the regents voted $15 a day for Captain Lea to take his horses and buggy to transport Meadors on a recruiting trip around Lincoln County and the mountains. Many other routine actions were taken to ensure that the Military Institute would be well prepared for the beginning of a new school year in the fall of 1898. Further, the Legislature added more tax support; and Congress finally gave NMMI 50,000 acres of land for its support in the Act of June 21, 1898.

The three-story building was completed in March 1898, just south of the present NMMI chapel and east of the present commandant's and superintendent's residences, according to Gibbs & Jackman. They named it 'Lea Hall' in honor of Captain J.C. Lea, the principal mover-and-shaker behind the establishment of the school. One building housed classrooms, kitchen and mess hall, offices, dormitories, and staff quarters.

A very important event in the history of NMMI was the hiring of James W. Willson as commandant of cadets. Dow explains some of the major changes brought about by Willson:

> From the very first, Colonel Willson made serious objections to the admission of day students, for he foresaw that they would be a constant source of trouble in a school under military discipline. The admission of girls to a school of this kind was positively too much for a military man to tolerate. Colonel Willson's arguments, however, were successful and at the end of the first session he secured an order from the Board of Regents refusing to admit day students and cutting off the Girls' Department entirely.

Actually, according to Gibbs & Jackman, while Willson was a "military school man" he had not actually served in the military. Dow remarked that these changes ended the military careers of several girls, including his future wife, Ella Lea (although her name does not appear on the roster of students for 1898-99). In addition to disallowing day students and girls, the regents limited enrollment to boys between the ages of fifteen and twenty years.

On September 6, 1898, New Mexico Military Institute began a new era in its history when the first term in the new building and in the new location started with appropriate ceremonies. Besides the student body and faculty, some two hundred citizens attended the function. John W. Poe was president of the board of regents at the time and, according to Kelly, Poe said, "On behalf of the board of regents and the territory of New Mexico, I now declare the New Mexico Military Institute formally opened."

Not everyone agreed that the renewed NMMI was, in fact, the same school that had previously operated at Seventh and Main Streets for one year and at Fifth and Main Streets for three years before that. Some accounts, yearbooks, and other sources reckon the beginning of NMMI from 1898, when it opened in the new building on North Hill. Others count the beginning from 1893, when the Territory of New Mexico officially adopted (but did not support) the little military school. It seems reasonable to assume that NMMI on The Hill is the descendant of Goss Military Institute that started in 1891 at Fifth and Main.

Wildy Lea attended Austin College in Sherman, Texas, and played sports there while NMMI was closed; but he came back to attend the military school in Roswell at least through the first year on North Hill. Baseball was the preferred sport on campus, and Wildy was on the team in the spring of 1899.

The growth and development of NMMI was largely routine after it was resurrected on North Hill. Various buildings and other facilities were added from time to time as needed because of increasing enrollment. A residence was finished by the beginning of 1899, described by Kelly as a ten-room cottage to house the superintendent and servants. Kelly states that at the end of 1899,

Above: George Curry helped get the legislation for Territorial take-over of Goss Military Institute. *Courtesy: HCSNM, Photo No. 3356*
Below: James F. Hinkle, manager of the CA Bar Ranch, carried the NMMI bill through the Legislature. He is shown here with his family in 1910. l. to r.: Hinkle, Vera, Rolla. Front row: Lillian, Clarence, Mrs. Hinkle. *Courtesy: HCSNM, Photo No. 2448*

another five-room cottage was built for the superintendent, leaving the ten-room house as a dormitory for cadets. In February 1900, Gibbs & Jackman relate, a 'crash program' was implemented for the immediate construction of a temporary, frame, five-room cottage for cadet quarters. According to Kelly, a new dormitory with room for up to thirty cadets was built in the summer of 1900.

In March 1901, Superintendent Meadors resigned, and the regents appointed James Willson to succeed him. Few individuals would have the amount of influence exercised by Willson over the history of NMMI. Dow states:

> The ideas of Captain Joseph C. Lea became the dream and its accomplishment the life work of Colonel James W. Willson, who vizualized *(sic)* the Institute campus, not as cluttered with straggling, frame buildings and poorly equipped barracks, but as a group of magnificent buildings, all constructed of a certain type of architecture, and each conforming to the other in style and beauty. Colonel Willson put his dream on paper. He drew pencil sketches of the dream buildings. He extended them into preliminary plans. Being always backed by a sympathetic and far-seeing Board of Regents and by the citizens in the Roswell community, he brought into reality many of the buildings of which he had dreamed.

The first class to graduate from NMMI,1902, had three cadets. Commencement exercises had been held in previous years despite the fact that no students were ready to graduate.

On April 8, 1902, the Corps of Cadets published the first edition of the NMMI yearbook, *The Bronco,* which was dedicated to Captain Lea. Gibbs & Jackman say this was an indication of the high esteem in which the cadets held Lea. Surprised and flattered by the tribute, the proud Captain rushed a copy off to Texas for his wife to see. In a letter to Mabel (cited by Padgitt), Lea considered both their roles in the establishment of the school and where the credit was due for getting it started:

> I was so surprised to find a little book or magazine dedicated to me by the school with my picture and a little

> biographical sketch of my life, which I have not read yet. I only saw the picture and headlines and immediately mailed it to you.
>
> It was certainly a surprise to me that anyone here would give me any credit, now since the school has got to be a credit to the town and all of New Mexico is proud of the Institution. Many others are clamoring for the credit of it. But you are the one who done the most to start it and to get Goss to come to Roswell and lay the foundation. It is certainly true that I got the legislature to pass the bill making it a Territorial Institution. That is sure. And it is further true that the school was at that time dead beyond resurrection and never would have been anything from that time on if it had not been made a Territorial school and placed under Territorial auspices.
>
> It is now acknowledged to be the leading school of all New Mexico. It has now got so far along that it will rank right up with the very best in all the southwest in a very few years, unless some very bad management occurs.

Another honor for Lea took place in September 1902 when some of the cadets organized the 'Lea Literary Society' for the purpose of becoming proficient in debating and public speaking. The Roosevelt Society, a similar organization, provided the competition, according to the yearbook.

Lea continued to serve on the Board of Regents up to the time of his death on February 4, 1904. According to Gibbs & Jackman, he was proclaimed as the 'Father of the Institute' in Special Order No. 36, issued on the day he died. The Corps of Cadets participated in Captain Lea's funeral, escorting the body to the South Side Cemetery. The colors were draped at the school from the time of Lea's death until the time of his burial, and the cadets wore 'mourning badges' for a month.

12

RAILROAD TO ROSWELL, 1894

Even before he began buying land from the Lea Cattle Company, J.J. Hagerman was already involved in the economic development of the Pecos Valley. His efforts exceeded those of Lea in some ways.

Hagerman, a native of Ontario, made a fortune in the iron industry in Michigan. Poor health required him to retire at the age of forty-four years, and he and his family moved to Colorado Springs. Hagerman regained his health and started investing in a railroad and in the mining of coal, silver, and gold. He soon had another fortune, according to John J. Lipsey in *The Lives of James John Hagerman* (1968).

Pat Garrett, Charles Green, and Charles B. Eddy started the Pecos Valley Irrigation and Investment Company in 1888 for the purpose of carrying out Garrett's idea of large-scale development of the Pecos Valley as a prosperous farming area. Eddy went to Colorado Springs to seek funding for the project, met Hagerman, and thereby started Hagerman's involvement with the development of the Pecos Valley.

When Hagerman became associated with Eddy and his partners in 1890, the corporation was reorganized as the 'Pecos Irrigation and Improvement Company' (PI&I) with Hagerman as the new president. Garrett was squeezed out. The company planned to construct dams, reservoirs, and canals. Hagerman decided that a railroad would also be needed if the agricultural potential of the Pecos Valley were to be realized, Richard H. Gravel states in his 1954 master's thesis (Eastern New Mexico University), 'The Pecos Valley Railroad, 1889-1906.'

Hagerman, Eddy, and eight others formed a Texas corporation on March 1, 1890, called 'Pecos River Railroad Company,' to build the fifty-four mile segment of the railroad from Pecos City, Texas, to the New Mexico Territory line. Hagerman set up a New Mexico corporation, 'Pecos Valley Railway Company,' on August 27, 1890, to put in the thirty-five miles from the boundary line to Eddy. Both companies were called 'PVR' or 'Pea Vine.' (See the present writer's article, "J.J. Hagerman and the Pecos River Railroad," in *The Permian Historical Annual*, 1978).

The first work train arrived in Eddy on January 10, 1891; and the town held a big celebration three days later, complete with fireworks. The festivities centered around the Hotel Hagerman, which Hagerman had built for the convenience of prospective settlers. Hagerman's private car arrived with railroad dignitaries.

Financing problems pretty much stalled the railroad at Eddy for some three years. The Panic of 1893, together with disastrous flooding of the Pecos Valley, slowed the extension of the road to Roswell. Claiming to be farther from a railroad than any other town in America, Roswell was connected to the outside world only by hack lines and freight wagons. Captain Lea and the leadership of Roswell endured frustrating times while waiting for the railroad to come to their train-starved town.

The *Roswell Record* of May 12, 1893, reported that surveying of the proposed route from Eddy to Roswell was completed and the location for the Roswell depot was chosen. By February 1894, no actual work had been done, so the Roswell leadership decided to take action. They solicited landowners, businesses, and professional men to pledge toward a $50,000 bonus that they offered to the PVR if the company could get the railroad to Roswell by September 1st. Captain Lea, his wife Mabel, and her brother William H. Doss, and many of Lea's other relatives and partners, were among over sixty donors who pledged either land or money:

Joseph C. Lea, Mabel Lea, W. H. Doss	100 Acres
William H. Cosgrove	40 Acres
George T. Davis	20 Acres
G.A. Richardson	1 lot
John W. Poe	40 Acres

J. Smith Lea	40 Acres
Joe D. Lea	30 Acres
Frank H. Lea	40 Acres
Ella Lea Calfee Pierce	10 Acres
Ernest L. Wildy	30 lots

The Roswell bonus, together with a tax exemption for new railroads, helped Hagerman acquire funding for the road. Work started, but the contractor, Joe Hampton, had built the road only to Miller (present Artesia) by September 2. Five weeks behind schedule, the first 'official' train (besides the work train) finally reached Roswell on October 6. The exuberant town fathers gave the $50,000 to the delinquent railroad anyway.

The Pecos Valley Railway locomotive's whistle announced to Roswell folks that the first official 'passenger' train had arrived on that late Saturday afternoon. The *Roswell Record* of October 12, 1894, reported that the news spread over the town in a few minutes that the special car of PVR General Manager Jeff N. Miller was out on the hill by the old school house. "There was a stampede for the scene of action headed by Capt. [J.C.] and Judge [Frank] Lea, driving Dick. When they got there the Capt. was joyfully surprised to meet Mrs. Lea, who came up on the special."

The engine, 'Roswell,' was bedecked with flags and golden eagles. The *Record* stated "... it was with admiration and unalloyed pleasure that our people looked upon that beautiful piece of machinery that is to play so important a part in the future in the development of this part of the Valley."

The train did not stay very long on its initial appearance because at 5:00 p.m. it had to move out south of Roswell to Zuber Hollow to a switch in order to allow the work train to pass. The Leas and many other Roswell families accepted Miller's invitation to join him for the short ride, so they became the first to ride a train out of Roswell. Dills, editor of the *Roswell Record*, and his family were among the riders. He wrote, "(The editor) took his station on the platform and looked over the road; ... we have never seen a road as well built and smooth as this road."

The first train arrives in Roswell on October 6, 1894, and much of the town turns out to greet the railroad dignitaries. The hatless man in the center is Joe D. Lea, first publisher of the *Roswell Record.*

Courtesy: HCSNM, Photo No. 311

When the train steamed back into town three hours later, the people were gathered. A crowd estimated at half the population of Roswell showed up at the track "... . and gave General Manager Miller and General Attorney Hawkins a rousing welcome." The town's brass band played; fireworks were set off; and, "... after much solicitation on the part of the populace," Hawkins made a brief talk to thank the people for their interest in the extension of the railroad into Roswell from the south. Miller declined to address the throng.

Later, the railroad guests were "... taken in hand by the members of the Roswell Club and entertained until the bedtime hour in their rooms," Dills wrote. He concluded, "Last Saturday night will long be remembered in Roswell, and those memories will be pleasant."

Roswell people had been busy for weeks planning 'Railroad Day' in Roswell to celebrate the great event. On Monday, October 15, they threw an all-day barbecue at Hagerman's Southspring Ranch (formerly John Chisum's South Spring River Ranch) southeast of town. About four thousand people attended, brought in by three special runs of the train from Roswell out to the South Spring River station at the ranch. Another train came up from Eddy.

The *Roswell Record* of October 19 reported that the people "entertained themselves by ... strolling through the grounds and viewing the incomparable loads of fruit that weighted down the orchard." The crowd ate twenty barbecued steers and five hundred watermelons in the shade of the big trees. The Roswell Fire Department served lemonade and ice cream to the happy celebrants.

After everyone was fed, the speeches started. Several spoke, including Governor W.T. Thornton, Congressional Delegate Antonio Joseph, and Parker Earle of Roswell. The highlight of the day was a speech by Hagerman himself, in which he spoke of the bright future he saw for Roswell and the Pecos Valley. He ended by thanking those present "... for your kind and hearty reception on this occasion."

That evening the Roswell Club hosted a ball to continue the celebration. Governor and Mrs. Thornton led the two hundred couples who responded when the grand march was called. Summing up the event, Dills accurately noted that "it is, and will continue to be, an epoch in the history of Roswell and Chaves county."

It would be difficult to overestimate the importance to the Pecos Valley of the completion of the railroad to Roswell. With this new development, Roswell residents could travel and ship their livestock and farm produce to other parts of the country. Together with related developments, it had a great economic impact on Southeast New Mexico. It was a fulfillment of J.C. Lea's vision for the development of the Pecos Valley, and a considerable boon to the land development program of the Lea Cattle Company. Agriculture grew in importance as many new farmers settled in the valley. Ranchers built loading pens along the railroad to ship out their livestock. New towns were started, including Hagerman. Because of the courage and foresight of Lea, Hagerman, and others who were involved in the early development of the Pecos Valley, Southeast New Mexico finally began to benefit from rail connections with the world.

As significant as the railroad was, however, there were still transportation problems for Pecos Valley producers. It was frustrating to have to ship cattle and fruit by way of Texas and the South when they were bound for northern markets. According to Keleher in *The Fabulous Frontier* (1962), the PVR was not profitable and was often criticized because "... it began nowhere and ended nowhere."

The PVR went into receivership in 1896, Lipsey states. What was needed, Hagerman allowed, was a northern connection for the railroad. When he started out to restructure the company, Albuquerque, Las Vegas, and Socorro all vied to be the northern terminus of the railroad. In the end, however, the PVR was built to Amarillo to connect with the Santa Fe Railroad. Hagerman formed the Pecos Valley and Northeastern Railway Company (PV&NE) in June 1897 to operate the extended line. In January 1898, the Santa Fe, Pullman, and several eastern financiers came up with the needed funding for the new 208-mile extension. The work of building the new road started in May 1898 in Amarillo and a few months later in Roswell; and the extension was completed on February 11, 1899, about where the railroad crosses the Pecos River some seventeen miles northeast of Roswell.

The economy finally recovered from the effects of the Panic of 1893. Despite its own financial problems, the PV&NE ultimately contributed to the prosperity of the Pecos Valley.

13

CHANGING THE CHARACTER OF ROSWELL'S GOVERNMENT, 1903

As the population of Roswell and Chaves County increased, local citizens began to ponder the advantages of re-organizing Roswell's town government to that of a city so the city could have more control over its own affairs. The town showed a population of 2,049 in the 1900 census, and the county had about 6,700. In the spring of 1902, the Board of Town Trustees instructed W.W. Gatewood, the town's attorney, to look into the issue of re-incorporating Roswell as a city.

Gatewood reported on June 17, 1902, that to be a city a town must have a minimum population of three thousand. Roswell obviously did not meet the population requirement, so the issue lay dormant for about a year.

The 1903 Legislature passed a new law which lowered the population requirement and provided that towns with two thousand or more people could be proclaimed a city by the governor of the Territory. The Roswell Town Board acted on the new law at their regular meeting of July 23, 1903, and passed Ordinance Number 182, instructing the town attorney to begin the preliminary steps to reincorporate the town as a city. Not only did they want to find out how many people lived within the corporate limits of the town, but they also proposed to annex some settled areas just outside the town boundaries. The ordinance described the proposed new boundaries and appointed W.C. Brass to take an official census of the town and of the proposed additions.

Brass made his report to the Board on September 4, 1903: there were 3,237 souls living within the town limits, plus another 1,303 people living in the adjacent areas that the council proposed to annex. This would give the new city of Roswell a population of 4,540—more than double the 2,049 in the 1900 census just three years earlier.

With the census information in hand, the board proceeded to instruct the town clerk to make application to the governor for the proclamation. Included with the application was a plat of the proposed city, a description of the center of town (the intersection of Fifth and Main Streets) and the boundary lines, and the assessed valuation of all property ($914,849) within the proposed city limits. Governor Miguel A. Otero received the application materials, ascertained that it was all in order, and issued his Proclamation. The heart of the Proclamation reads as follows:

> Now, Therefore, I, Miguel A. Otero, Governor of the Territory of New Mexico, by virtue of the authority in me vested by the laws of said Territory, do proclaim and declare:
>
> That under and by virtue of the provisions of Chapter 111 of the laws of the Thirty-fifth Legislative Assembly of the Territory of New Mexico, entitled "An act to enable towns or villages in the Territory of New Mexico to change the character of their municipal government, and for other purposes" approved March __, 1903, the town of Roswell is entitled to become a city.
>
> That the name of such city shall be The City of Roswell….
>
> … and the boundary lines herein described shall be the boundaries of such city and all the territory included therein shall constitute and be known as the City of Roswell, and shall be governed as such by the officers thereof, under the laws applicable thereto.
>
> Done at the executive office this the 25th day of September, A. D. 1903. Witness my hand and the great seal of the Territory of New Mexico.
>
> MIGUEL A. OTERO.
>
> By the Governor.
> (Great Seal)
> J. W. RAYNOLDS.
> Secretary of New Mexico.

The boundaries are described in the document in terms of townships, ranges, and sections; but an approximation of the city limits in terms of streets is possible. Some of the present-day streets had not been established or named at the time. Generally, it can be stated that the boundaries of the new City of Roswell as it was established in 1903 were as follows:

> East boundary: South and North Atkinson Avenue from East McGaffey Street northward to East Nineteenth Street.
>
> North boundary: East and West Nineteenth Street from North Atkinson Avenue westward to North Montana Avenue.
>
> West boundary: North Montana Avenue southward from West Nineteenth Street to West Eighth Street, jog west on West Eighth to North Wyoming Avenue, south to Riverside Drive, east back to Montana, then continue south on Montana and Sunset Avenue to about a half-block north of where West Deming Street intersects South Sunset Avenue.
>
> South boundary: eastward from South Sunset Avenue along a line half-way between West Deming and West Albuquerque Streets to South Washington Avenue, then south on Washington to West Chisum Street, east on West and East Chisum to South Cahoon Avenue, north on Cahoon to East McGaffey Street, and finally east back to the starting point on South Atkinson Avenue.

With the governor's proclamation accomplished, on October 29, 1903, the City Council called a special election to be held on December 8. The purpose as stated in the call was to elect a mayor, city clerk (sometimes called 'recorder'), and city treasurer from the city at large, one alderman from each ward, and two members of the board of education from each ward. Polling places for the respective wards were established at the county courthouse, E.B. Stone's store, Central School, Pauly School, and Charles Whiteman & Brothers' store. Registration was required, and a total of 1,018 voters who lived in the expanded city limits registered.

The Board of Town Trustees prior to the changes of 1903 was non-partisan. While most people around Roswell considered themselves to be Democrats, the out-going chairman and clerk of the town board, Nathan Jaffa and Robert Kellahin, respectively, were both Republicans. They, and many other leading citizens, believed that city government was more business than politics and therefore it need not be based on partisan politics. There is no Democratic way or Republican way, for example, to sprinkle water on dirt streets to hold down the dust.

Some of the Democrats of Roswell, however, saw an opportunity in the restructuring of the city government to elect members of their party to office. Some of the Democratic leaders got together at the Chamber of Commerce on November 11 and decided to organize the party in the city and in the wards, with the purpose of electing Democrats to city offices. They issued a request to the wards to hold meetings on November 17 to each nominate one candidate for alderman, two for the school board, and five delegates to a city convention.

The partisan activity of some Democrats disturbed others who were Republicans, independents, or Democrats who were opposed to party politics in city government. Later, the *Roswell Register*, a Republican paper, characterized the Democratic move as "... the few Democrats, who in a secret meeting arranged their little slate ... ." The newspaper predicted the reorganization of the city government would not take place without dissent. A group of thirty-eight "energetic and wide-awake citizens" composed of members of both political parties issued a call for a Citizens' convention. The call was published in the *Roswell Daily Record* on November 14, 1903, announcing a meeting at the opera house on the 16th. The statement reads, in part, "We the undersigned do not deem it a wise policy for the best interest of Roswell to have any class or party ticket in the coming town election."

John W. Poe was elected chairman of the convention and C.C. Tannehill, secretary; they were both Democrats. The convention voted to nominate candidates to run as independents for the at-large positions of mayor, clerk, and treasurer, and to call for ward caucuses later in the week to nominate aldermen. Delegates nominated three men for mayoral candidate: Captain J.C. Lea, Judge Frank H. Lea, and Nathan Jaffa. By ballot the convention chose Nathan Jaffa, Republican, to run for mayor and Robert Kellahin, Republican, for recorder. E.H. Williams, a Democrat who was not present, was nominated for treasurer. The *Register* openly supported Kellahin.

The partisan Democrats were not to be deterred. The ward caucuses on November 17 were large and enthusiastic, and some of them carried on until a late hour. The *Daily Record* could not resist some good-natured ribbing of the Fourth Ward, which met at the 'Pauly Addition school house.' The school was on the west side of South Union Avenue between present West Tilden Street (Santa Fe Street at the time) and present West Hendricks Street (Las Vegas Street at the time). "The place of meeting was so far out that it is said many who started got lost and are yet wandering around in the vicinity of the Capitan mountains." The Fourth Ward also had the distinction of being the only ward that was not unanimous in their selection of candidates and delegates.

A story in Roswell history is that Captain J.C. Lea volunteered to be the first mayor and serve until the regular election cycle came around. The evidence suggests that in fact he was drafted. All of the wards except the Fourth Ward instructed their delegates to vote for Lea for the Democratic nomination for mayor. Those from the Fourth were all in favor of Lea but not instructed.

The *Roswell Daily Record* of November 18, reported:

> Captain J.C. Lea, the "man who blazed the way," and on whose homestead Roswell is built, will Saturday night receive the unanimous nomination for the first mayor of the city, built from the town in which he has lived for so long and loved so well.

The five ward caucuses chose as alderman candidates L.B. Tannehill, Ralph M. Parsons, Samuel P. Denning, W.W. Ogle, and A.L. Whiteman, respectively.

The Citizens' movement to counter the powerful Democratic effort fell apart. Apparently the ward caucuses were not held, and some of the Democrats decided to go with their party instead of with the independents. Whether or not the certain nomination of Captain Lea and others for offices in the new city government by the Democrats caused the Citizens to reconsider is not known.

Their mayoral candidate, the incumbent chairman of the Board of Town Trustees, Nathan Jaffa, published a letter in the *Roswell Daily Record* on November 19 in which he explained why he could not continue as a candidate for mayor. He saw it all as a Democratic move to keep Republicans out of office:

> Not deeming conditions such as to warrant this antagonism, I deem it expedient at this time to withdraw my name as candidate for mayor on the Citizens' ticket .... I have always thought it unwise to instill politics into our town affairs and have not changed my opinion in the least, believing that the upbuilding of Roswell required the combined efforts of all parties regardless of politics, and the present scramble for office is to my mind unjustifiable and hurtful to the best interests of our city. I therefore decline to engage in it.

The city convention of the Democratic Party of Roswell at the Chaves County Courthouse on Saturday evening, November 21, 1903, was viewed as a major affair. It was the first such convention and marked the organization of the party as a city party. The meeting was touted as a rally as well as a convention, and a special invitation was issued to ladies to attend. There were speeches and music, followed by the ratification of the aldermen's nominations and the official nomination of the three citywide candidates. The convention also chose a central party committee to be in charge of running the campaign.

When convention chairman G.A. Richardson called for the nominations for mayor, no less a personage than the former Populist Congressman from Kansas, 'Sockless' Jerry Simpson, "... in a speech of highest eulogy," placed the name of Captain Lea in nomination for mayor. A standing vote amid deafening applause made the choice unanimous.

The crowd called for Lea to make a speech, and some of his remarks were carried in the *Roswell Record* of November 27:

> Gentlemen of the convention, for this honor I thank you more than words can tell. I am not a public speaker, so I cannot make you a speech. Still I ought to be full of speeches, for no

Above: G.A. Richardson, the first attorney in Roswell, played a role in the reorganization of Roswell's government, was involved in the Lea Cattle Company, and in the creation of NMMI.
Courtesy: HCSNM, Photo No. 4009
Below: Roswell just after the turn of the 20th century. View is south from Fourth St.
Courtesy: HCSNM, Photo No. 1819-B

this honor which you have conferred on me. I am glad to accept it at your hands. My heart is too full to express to you my feelings. Gentlemen, I thank you.

E.H. Williams, who had been nominated by the Citizens' ticket for treasurer, declined that nomination but accepted nomination for the same office by the Democrats. Fred J. Beck was nominated for clerk.

The special election of December 8, 1903, turned out to be a non-event, as only 460 of the 1,018 registered voters chose to exercise the franchise on that day. The low turnout was blamed on the fact that there was no competition for most of the offices. Most Democrats did not bother to vote, although virtually all of the few Republicans voted. The outcome was to be expected: all of the Democratic candidates were elected, even in the contested races.

Captain J.C. Lea received 443 of the total 460 votes cast. Some voters did not vote for all the positions, but nobody voted against Lea. In the Third Ward, someone wrote in the name of Frank Blashek, the miller; but that was the only vote for mayor besides those for Lea. The *Roswell Record* allowed that "... it goes down in the history of the first city election of Roswell to the great credit of [those] who voted that a pencil mark did not scar the name of Captain J.C. Lea."

Robert Kellahin, who had been the town clerk for several years, ran as an independent on the Citizens' ticket against the Democratic candidate, Fred J. Beck. Beck narrowly won the contest by a vote of 231 to 229. In the Fourth Ward, the labor unions ran A.W. McWhirt for alderman; but the Democratic candidate, W.W. Ogle, defeated him thirty-three to twenty-four.

The *Record* concluded, "Thus Roswell's first city administration will be in [the] control of the Democratic party, that must henceforth be held responsible for its good government." Roswell was on its way to becoming one of the largest and most important cities in New Mexico.

The new City Council met for the first time on December 21, 1903. The new mayor and aldermen took the oath of office from

Justice-of-the-Peace J.M. Peacock. Mayor Lea then adjourned the council until two days later, at which time he was ready to make his appointments. The council voted to approve each appointment by a roll-call vote. Among those chosen were: City Attorney, G.A. Richardson; Engineer, W.A. Wilson; Physician, Dr. W.W. Phillips; President of the Council, Alderman L.B. Tannehill; and Fire Chief, Alderman R.M. Parsons. Mayor Lea appointed the aldermen to the following standing committees: Streets and Alleys: Tannehill and Denning; Finances: Parsons and Ogle; Ordinances: Tannehill and Ogle; Poor and Paupers: Parsons, Denning, Whiteman (and the mayor?); and Sanitary: Parsons, Denning, Ogle, Whiteman.

The December 21 and 23 meetings were the only city council sessions over which Mayor Lea presided. Tannehill, as president of the council, conducted the meetings. In the regular meeting on February 2, 1904, the council, "... being advised of the serious illness of Mayor Lea and of his inability to perform the duties of Mayor of the City ...," proceeded to appoint Tannehill to "... act as Mayor of the City of Roswell ... until the further action of the council."

14

END OF THE TRAIL

Joseph C. Lea lived to see many of his dreams come to pass, but he had his share of disappointments and sorrows along the way. He enjoyed an active life, being involved in the cattle business, the real estate business, the New Mexico Military Institute, the City of Roswell, the Masonic Lodge, the Democratic Party, the Confederate Veterans, his family, and numerous other activities and issues that arose from time to time. He saw both of his children marry but may not have lived to see his first grandchild.

Wildy Lea

Wildy Lea attended New Mexico Military Institute during the 1898-1899 school year, by which time he was twenty-two years old. After that, few memories of his adult life have survived. He married Alice Howard of Hereford, Texas, about 1901 when he was twenty-four years old. They spent their honeymoon at the Montezuma Hotel near Las Vegas, New Mexico. They lived in Hereford, where their only child, Annie Wildy Lea, was born December 7, but the year is uncertain: 1902, 1903, or 1904, according to Thomas Lea Garry of San Gabriel, California who is Annie's son.

About the time of his marriage and before the birth of his daughter, Wildy Lea began to experience mental problems. According to Garry, Lea suffered a head injury in a rodeo shortly before he and Alice were married. The injury made it necessary for surgeons to implant a silver plate in Lea's skull, Garry said. As a result, Lea experienced attacks of paranoia and had to be institutionalized. Ella Lea Dow wrote in her notes, "Wildy Lea thrown from a horse, skull fractured. Las Vegas." She told her grandchildren that her brother was thrown or kicked in the head by a horse, which caused him to lose his sanity, according to Ella's granddaughter, Lea Rohr Fraser of Roswell. Ms. Fraser has reservations about the veracity of this story, which she believes could be a defense mechanism devised by Ella to soften the stigma of Wildy's mental problems.

A 1903 legal document connected with the case of Nate Hendricks, accused of killing Deputy Sheriff Will Rainbolt in 1901, has some information about Lea's plight. Hendricks was a friend of Lea and wanted Lea to testify in his behalf at his 1903 trial. However, Lea was "under temporary mental derangement, and is being treated for the same in the asylum at Las Vegas, New Mexico." The document states that Lea had this problem once before but "... his recovery from said previous temporary mental derangement seemed to be complete and he thereafter continued in an apparently normal and sane state of mind for about six months ... ." The document avers that ordinarily Lea "... is a man of more than ordinary intelligence, acuteness of mind and of excellent memory ... it is believed and here stated that the present attack of said Lea is only temporary ... ." Hendricks' document expresses confidence that Lea would "be shortly fully and permanently restored to a sound state of mind; that said Lea is a man of personal integrity and of a high sense of honor and one whose reputation from truth and veracity is above reproach ... ."

A letter written by Captain Lea to his wife in Texas on October 1, 1902 (in his probate records), reveals how he felt about his son's illness:

> I think Wildy is certainly much better. I will send you his last letter & the Dr's. last letter. He writes that he is now satisfied that he will make a permanent cure of Wildy, but he wants him to stay there several months yet.
>
> I don't want Wildy to ever come here to live again when he gets well. I want him to go to Canada or Montana or Mexico or Texas or some place away from here. I am satisfied that if he gets well that he will make a success of cattle & a farm or ranch. Capt. Willingham wants him to go with him to Canada to Montana & go into the cattle business there. He asked me to let him take Wildy with him when he gets well. His son Drew

> has entirely recovered so he tells me & is making a fine man on his ranch in the Panhandle, & is now a big strong man & has all of his proper senses about him ... . Please write Wildy a cheerful encouraging letter, to stay there until the Dr. says he is cured. I want him to stay until Dr. Tipton says he is cured ... .

Unfortunately, Wildy Lea spent the rest of his life in what at that time was called the "New Mexico Territorial Insane Asylum" in Las Vegas, classified as a 'lunatic.' A committee of relatives and close friends handled his business interests in Roswell.

Ella Lea

Ella Lea was a tomboy who could beat the boys in climbing trees, according to George Gerling in a chapter in *Roundup on the Pecos* (1978). In fact, she had her own apple tree in the family orchard that nobody else was allowed to climb. A childhood friend of Ella was Flora Whiteman (Miller), who said Ella started wearing stylish clothing early in life. "When she came back from living with the Thurbers in California [1889], she wore gloves." She always wore the latest fashions, custom made for her by dressmakers; but she thought nothing of wearing velvet for everyday wear.

Ella attended Goss Military Institute and New Mexico Military Institute for a few years in their early days. In due time, Ella went to the Kidd-Key Boarding School in Sherman, Texas, which was the fashionable finishing school of the day for socialite girls. She afterward attended Christian College in Columbia, Missouri. Her education was greatly enhanced by her love of reading books.

Ella became an excellent horsewoman. Her favorite activities were riding sidesaddle on her horse of choice, 'Lady,' and driving a carriage. She often rode Lady in local parades, decked out in a plumed hat and a sweeping gown.

The *Roswell Record* of October 30, 1903, called Ella "... one of Roswell's most beautiful and charming society ladies." She was also very eligible and available for marriage to some fortunate young man. That man turned out to be Edgar Lucius 'Ted' Bedell, who was twenty-five years old when he arrived in Roswell in 1900. In her notes, Ella says he was born in Chetopa, Kansas, on

Above: Ella Lea, age 18, on August 12, 1899. Ella was one of the most beautiful and eligible young women in Roswell.
Courtesy: HCSNM, Photo No. 1893
Below: Ella Lea in her buggy, driving her favorite horse 'Lady.'
Courtesy: HCSNM, Photo No. 625

August 8, 1875; but Gerling, who got his information from Ella's daughter Josephine Bedell Rohr, says Bedell was a young journalist from New York City. He went to Kansas to find a career and a better climate for his health. He became acquainted with the famous editor of the *Emporia Gazette*, William Allen White, who inspired him to pursue journalism. When Bedell arrived in Roswell, he was unable to find a position as a journalist. He took a job at the Citizen's National Bank and soon became the assistant cashier.

Bedell and Miss Lea were attracted to each other and soon planned to be married. The wedding was not a large affair. The *Record* headline stated, "Roswell society is pleasantly surprised by marriage of popular young couple." The "impressive ceremony" took place in the rectory of St. Andrew's Episcopal Church on the evening of October 28, 1903. The Reverend George Hinson, Rector, officiated; and the witnesses were Dr. Bishop, Maude Keller, and Lillian McCain. There is no mention in the newspaper account about whether or not her father, stepmother, or any other relatives were present. The *Record* reported on October 30, "Mr. and Mrs. Bedell are now at the Grand Central where they are receiving the congratulations of their many friends. The groom is a very popular young man ... ." Captain Lea did not live to see Ella and Ted Bedell's two daughters who were both born after Lea died.

Hot Springs Natatorium

One of Captain Lea's last projects, besides being involved in the reorganization of the town into a city, was to build a hot springs natatorium and park in the center of the city. The editor of the *Roswell Daily Record*, Harvey F. M. Bear, spoke frequently with Lea about the project and wrote about it in the paper after Lea died. Lea firmly believed Roswell was destined to be the major metropolis of New Mexico, and all of his plans were made with this in mind. Around 1903, the natatorium was one of his most cherished plans. He wanted a deep well drilled in the very heart of the city, much deeper than any that had ever been drilled before, reaching down many hundreds of feet to the rich, warm waters that he believed nature may have hidden in reservoirs below the city.

Around this well he proposed to build a public park with a modern natatorium in the center. Lea believed the cost of this project would be relatively small, and it would be one of the best things Roswell could do. Bear writes:

> Not only would this be a boon to the heath seeker, a feature of beauty and utility to the city, but at the same time one of great commercial value. The deep waters below us have never been tapped and we do not know what treasures may be stored there for us. The natatorium would not only be a matter of comfort for our own people but would be the means of bringing many here and in many ways would be of great advantage

According to Bear, this was not an idle dream with Captain Lea, who was known for transforming his dreams into reality. Lea set about securing subscriptions from his friends and fellow citizens to finance his spa project. At the time of his death, he had pledges sufficient to cover more than two-thirds of the anticipated cost of the project. Unfortunately, the project was dropped because of Lea's death.

Death of Captain Lea

On January 28, 1905, Captain Lea became ill and soon developed pneumonia. Dr. W.T. Joyner was his physician; and he visited Lea at his home every day during his fatal illness—sometimes several times each day. Dr. Joyner was there five times on February 2, when the prognosis became grim; and he brought in his partner, Dr. C.M. Yater, to consult with him on the case. On February 4, as stated in the flowery language typical of newspapers of the time, "... just as the sun was at his zenith, the grim messenger came and his spirit took its flight to that bourne from whence no traveler returns." The old soldier was sixty-two years old.

During the next few days after Lea's death, the Roswell newspapers—the *Roswell Daily Record*, the weekly *Roswell Record*, and the *Roswell Register*—devoted much space to Captain's Lea's

Above: Captain J.C. Lea late in life.
Courtesy: HCSNM

Left: Captain J.C. Lea's grave, South Park Cemetery, Roswell.
Courtesy: HCSNM, Photo No. 1910

life, death, and funeral. Many of the articles are replete with lavish praise for the pioneer developer of the city. One example was printed in the February 12 edition of the *Record*:

> Capt. Lea was in almost every aspect a remarkable man. In stature he stood six feet and four inches and his nobility of nature was as far above that of the average man as he exceeded him in stature ... . [T]o his aid more than to any one or dozen is the growth of his beloved town due ... . [H]ad he been a selfish gain-seeker he could have been one of the wealthiest men in all the land, but such was the breadth of his charity that he died comparatively a poor man. No worthy person every applied to him in vain ... .

The following account is based largely on those contemporary news articles. The recognition accorded Lea and the funeral carried out by the community were commensurate with his great stature in the Pecos Valley and would have befitted even a personage of such immense importance as a governor or head of state. Shock ran through the Roswell community at the news of Lea's death. According to the newspaper accounts, every Roswell home felt a personal loss and grief because "this man was regarded by the people of Roswell with much of the same affection which children feel for a father." He truly was the 'Father of Roswell.'

Normal community activity stopped immediately. Two days earlier, the City Council had appointed Council President L.B. Tannehill to act as mayor until further action of the council. Upon learning of Lea's death on February 4, Tannehill called a special council meeting that same afternoon at the Chamber of Commerce office. They issued a proclamation ordering all city offices and buildings closed until after the funeral and burial of Mayor Lea, with "proper and usual emblems of mourning" placed upon them. Further, the Council requested that all businesses and the public schools close until after the burial. The schools carried out the request, although it is not known whether all businesses closed. It is probable they would have been normally closed anyway on Sunday, the day of the funeral.

On Friday morning the Valverde Camp No. 1419, United Confederate Veterans, met to pass a resolution of respect for their departed commander. The resolution also expressed "our tenderest sympathies and affections" to the family, especially to Lea's daughter Ella Bedell, who was the sponsor of the Camp. They made plans for participating in the funeral and passed a separate resolution inviting "all the old Union soldiers" to unite with the Confederate Veterans in the funeral services. Dr. Edwin H. Skipwith was chosen as temporary commander to be in charge of not only the Confederate Veterans but also the New Mexico Military Institute cadets who took part in the funeral and burial services.

The Masonic Lodge met on Saturday morning and passed a resolution to express their profound sorrow and regret over the loss of their "honored and useful brother," and they conveyed the expression of "sincere and heartfelt sympathy" to the family. The Masons made plans to assist in the funeral and burial services, which were under their auspices.

Captain Lea's body was laid out in his home—the old hotel that Van C. Smith had built around 1870. The United Confederate Veterans posted men who guarded the remains day and night until the funeral service was completed. Lea's relatives were notified of his passing, but none of his surviving brothers or sisters was able to come. His wife Mabel and her daughter Willie Day were on their way back to Roswell from Dallas, presumably having been notified of Lea's grave condition. When their train arrived on that Thursday afternoon, they learned Lea had died just hours before they reached Roswell.

Lea's body was at his home until a special $250 casket arrived by train from St. Louis on Saturday afternoon. At that time his body was placed in the casket and taken to First Christian Church at West Fourth Street and North Richardson Avenue, where Lea was a member. The body lay in state in the church from 6:00 p.m. on Saturday until the start of the funeral at 10:00 a.m. on Sunday, February 7, 1904. Lea's friends were invited to view the remains at the church during those hours, and many hundreds did so. The casket was not to be opened during the funeral except for viewing by relatives. Clarence Ullery, undertaker, of the Ullery Furniture Company, carried out the funeral arrangements.

The *Record* asserted that the funeral was "the largest and most imposing funeral ever held in the Territory." It started at the appointed hour, presided over by the Reverend B.F. Wilson, former pastor of the church. Pall bearers were William Atkinson, Edward A. Cahoon, Nathan Jaffa, Jason W. James, L.W. Neatherlin, John Stone, J.P. White, and Charles Wilson, all prominent men of the area who were charter members of the Roswell Masonic Lodge No. 18. Seating in the church was reserved for Lea's family and close relatives, members of the Roswell City Council, Fifth District Court officials, the Board of Regents of New Mexico Military Institute, and both Confederate and Union Civil War veterans. The building was too small to seat all of the invited guests. Hundreds of others gathered in the street outside in all four directions from the church. The district court officials and attorneys were due in court in Carlsbad on Monday morning, so the Pecos Valley & Northeastern Railway made arrangements for a special train to take them there on Sunday afternoon after the funeral.

Most, perhaps all, of the other churches dismissed their Sunday morning services so their members could honor the memory of Mayor Lea. Dr. C.E. Lukens, pastor of the First Presbyterian Church, and the Reverend W.E. Lyon, pastor of the First Methodist Episcopal Church, South, both assisted with the funeral. Music was provided by a ladies' duet and a mixed sextet.

At the close of the service the body was turned over to the Masons at the door of the church, and they led the procession to South Side Cemetery. The Grand Central Hotel and other parts of Main Street along the processional route were appropriately decorated. There were more than two hundred vehicles in the process, which required a half-hour to travel the three miles to the cemetery south of the city. The city marshal and his deputies directed the procession. Estimates were that over a thousand marched in the procession and hundreds of others went on ahead to the cemetery to await the arrival of the funeral party.

When the cortege reached the entrance to the cemetery on present South Main Street, the cadets of New Mexico Military Institute met it and escorted it to the gravesite. The cadets served as guard to the grave (South Masonic Circle, Lot Thirty-seven). Graveside services were conducted by the Masons under the leadership of John W. Poe, assisted by the Reverend Wilson and Colonel J.W. Willson of the Institute. The Confederate Veterans held the Stars-and-Bars flag over the grave while cadets fired a salute of three volleys over the grave. Cadel L.M. Brownell stepped forward and sounded "Taps" on his bugle. The ceremonies were over. The great man was committed to history.

Captain Lea's Estate

As soon as Captain Lea was dead and buried, the enmity that had long been building between his widow and his daughter surfaced. The situation became very ugly, and it greatly complicated the settlement of Lea's estate.

Probate Judge J.T. Evans appointed Mabel Lea as temporary administrator of the estate of her deceased husband, but Ella Lea Bedell and her brother's committee immediately petitioned to block Mabel's appointment. They asked the court to recognize a will written by Lea on May 16, 1887, in which he made his children the sole heirs. In her petition, Mrs. Bedell asserted:

> ... Mrs. Mabel Lea has an utter hatred and contempt for ... Ella Lea Bedell and has repeatedly expressed her intention to ruin her if she could and has given vent to her ill feelings to [Ella] ever since the death of the said Joseph C. Lea to numerous persons ... .
>
> Mabel Lea has exhibited an unwarranted disposition to intermeddle in the private affairs of ... Harry Wildy Lea, without authority, in that immediately upon the death of Joseph C. Lea ... she caused all tenants occupying buildings belonging absolutely to the said Harry Wildy Lea in his own right, not to pay any more rent to anyone but herself, when she well knew that the said Harry Wildy Lea was a lunatic, in the Territorial Insane Asylum, with no means of support and that his wife and child were depending upon his income for support.

Above: Ella Lea Bedell and Ted Bedell with their first daughter, Eleanor. The Dorothy Bedell on the left is possibly Ted's sister. *Courtesy: HCSNM, Photo No. 665-B*
Below: Residence of Captain J.C. Lea, the first building in Roswell. Captain Lea is in the buggy.
Photo by Frost's Views. Courtesy: HCSNM, Photo No. 2043

Mrs. Bedell further stated that Mabel Lea had all she could handle without taking on the job of administrator of J.C. Lea's estate, because she had her ranch and other businesses in Texas to see after and was already the administrator of her brother's rather large estate. Further, Mrs. Bedell maintained that Mabel Lea was not even a resident of New Mexico Territory and that she had not lived with J.C. Lea as his wife for several years:

> ... prior to the year 1896 (Mabel Lea) maintained her residence here and during the years prior to said year was absent the greater part of the time ... .
> ... for more than seven years prior to the death of Joseph C. Lea, the said Mrs. Mabel [Lea] lived in the City of Dallas and State of Texas and during all of said time did not live with the said Joseph C. Lea as his wife but lived separate and apart from him ... .

In her petition to the court in response to Mrs. Bedell's charges, Mrs. Lea answered most of the allegations. She stated that she had been a bonafide resident of Chaves County for some fifteen years—ever since she married J.C. Lea—and was "actually domicled" in the county and Territory. She asked the court to disallow the 1887 will as invalid, not only because it did not meet legal requirements but also because it did not take into account that she had an interest in community property real estate acquired by the Leas subsequent to their marriage in 1889.

Mrs. Lea further maintained that she was the best-suited person to administer the estate of her late husband because:

> ... she has had a large business experience extending over a period of 20 years ... during which time she has had the control and management of large and various business enterprises ... . She further represents that she is solvent, being worth in her own right ... more than one hundred thousand dollars ... .

> [U]p to the time of the death of the said deceased on the 4th day of February 1904, the relations existing between her said husband and herself were always such as should exist between a dutiful wife and a loving, affectionate and considerate husband,

Dr. Parkhurst and Dr. Joyner's office, 124 East First Street, *c.* 1902.
l. to *r.*: Frank H. Lea, J.C. Lea, Mary Lea Joyner, Pearl Lea Kenny, Mildred Joyner, Vatelle Dills, Gertie Lea Dills, and Dr. C.M. Yater.
Courtesy: HCSNM, Photo No. 372

> as will appear from many letters ... which ... show the high regard in which her said deceased husband held her business ability ... .

As to Mrs. Bedell's other charges, Mrs. Lea stated that her other duties would not interfere with her administration of Lea's estate. She also asserted that she had no feeling whatsoever that was antagonistic to the interests of Lea's children; conversely, she averred, she would be delighted and pleased to handle the estate in such a manner as to see that they received the greatest possible benefit to which they were entitled.

Mrs. Lea submitted some letters to prove her case, and they have been preserved in J.C. Lea's probate records (No. 113) at the Chaves County Courthouse. They were written by Lea to his wife in 1902 and 1903, and they actually prove both points of view to some extent. All of the letters begin, "My Dear Wife," and close with "Your aff[ectionate] loving and devoted husband" or similar expressions. He addresses her in the letter written in November 1903 as "my darling."

The letters contain financial advice and comments on various business deals, both his and hers. However, statements in the letters seem to confirm that Mrs. Lea was in fact living in Texas; and he was planning to wind up his affairs in Roswell and move to Texas to be with her and her daughter Willie:

> I am so glad to know that all of our sympathies are always with each other. I think that is so nice even if we can't be together yet ... .
>
> That house might not be a bad place for us to live in, for [San] Angelo is a good place to trade & that this is a good country & and a good town, when it gets more railroads ... .
>
> Please keep me well posted on your whereabouts & what you're doing ... .
>
> I am coming, if God spares me & just as quick as possible & I expect I will stay until you are both tired of me.
>
> After this I will visit in New Mexico. You know you always said I only visited in Texas. We will change now, for awhile at least.

> You don't know all of my troubles here for the last 5 years. I would never tell you half, & very little because I never wanted to make you more unhappy, when you were already & always in such great distress. But if God will let me live I will soon be with you & try to make you both happy .... I won't have to run back here to attend to business.

The letters are addressed to Mrs. Lea at Dallas, Temple, and Talpa, Texas—the last being the postoffice nearest to her Coleman County ranch.

Judge Evans ruled that the 1887 will was not valid, thus opening the way for Ella Bedell and her brother's committee to appoint an administrator. They appointed former Roswell Postmaster Lucius K. McGaffey.

McGaffey found that Lea's personal property was worth only about $500 and consisted mainly of six horses. McGaffey reported on November 12, 1904, that Lea's real estate was worth $14,878. Mrs. Lea filed a claim against her late husband's estate for $25,000, maintaining that she had an interest in community property and that Lea had illegally signed her name on some land deals as her attorney-in-fact. McGaffey filed an argument in which he maintains that "this trumped up claim of Mabel Lea" was not valid and that the statement "... bears out her much heralded reputation of being 'smart' ... ."

Mabel D. Lea died in Dallas on April 4, 1906, before the case was settled. Her daughter Willie Day Padgitt, now married to J. Thomas Padgitt, Sr., was named the executor of her estate. On August 31, 1907, Mrs. Padgitt agreed to accept $5,000 in exchange for giving up all claims to Lea's estate.

All of Lea's creditors were satisfied by the end of 1907. McGaffey filed his final report on July 5, 1910, more than six years after Lea's death. The bitterness involved in the settlement of Lea's estate was a rather sad conclusion to his remarkable life.

15

EPILOGUE

Even after his death, Captain Lea's family and his influence continued to affect developments in Southeast New Mexico. His relatives moved on with their lives. The cattle company Lea had started finally dissolved, but the military school and the city that Lea had worked so hard to develop both continued to advance.

Wildy Lea

Captain Lea's son apparently did not attend his father's funeral. It is not known whether he ever visited Roswell again after he was confined to the mental hospital in Las Vegas. Samuel T. Howard, Wildy Lea's father-in-law, had been appointed in Texas to take charge of Lea's affairs. Since Howard lived in Texas and Lea owned property in the Roswell area, there was a need to have a local committee in charge of Lea's affairs. The bond records of Chaves County show that Judge William H. Pope, on April 22, 1905, appointed J. Smith Lea, Wildy's first-cousin-once-removed, and Milo L. Pierce, Wildy's uncle by marriage to Ella Lea Calfee, to be the committee "... to take charge of the person and property of Wildy Lea."

Wildy Lea spent the rest of his life (about thirty-six years) in the mental hospital and died there at the age of sixty-one on March 19, 1938, of heart trouble. Pastor D.B. Titus of Roswell's First Christian Church conducted the funeral three days later at Ballard's chapel. They buried him in South Park Cemetery in Roswell between his parents' graves. Lea's widow, Alice Howard Lea, died in Denver later the same year.

Ella Lea Bedell Dow

Ella, Captain Lea's daughter, married Ted Bedell before her father died; but her children were born after his death. Eleanor Lea Bedell was born on October 25, 1904. Her sister, Josephine Lea Bedell, was born on October 2, 1910. Ella and Ted Bedell divorced in 1911. Bedell went to the Pacific Coast and became a journalist again in San Francisco, Seattle, and Anchorage. He died in Tulare, California, on March 25, 1928.

On August 8, 1913, Ella married Hiram M. Dow. Dow was born in Cotulla, Texas, on April 21, 1885. His father, Les Dow, became a famous sheriff of Eddy County, New Mexico. Hiram graduated from the New Mexico Military Institute in 1904 and received his law degree from Washington and Lee University in 1908. Dow served as Roswell's city attorney for a time and later as mayor. Hi and Ella's daughter, Dorothy Lea Dow, was born on May 24, 1914; and Dow raised the Bedell girls as his own.

In 1936, Dow was elected lieutenant governor on the Democratic ticket with Governor Clyde Tingley and served one term. Dow became a very distinguished attorney, especially for the petroleum industry. He served many years as Civilian Aide to the Secretary of the Army and received many other honors and recognitions during his long lifetime.

Ella Dow was active in the Roswell Woman's Club and the local chapter of the Daughters of the American Revolution (DAR). She was very proud of her admission to the DAR; she worked hard for a long time to document all of the needed information. The Dows lived for many years at 212 West Thirteenth Street near the campus of New Mexico Military Institute. Mrs. Dow, the first Anglo child born in Roswell, died of heart problems on May 22, 1962, at the age of eighty-one. Her funeral was two days later at Ballard's chapel, where the Reverend Charles R. Davies of St. Andrew's Episcopal Church officiated. Hi Dow lived about seven years after his wife died. He passed away at the age of eighty-four.

Death of Frank H. Lea

Captain Lea's young brother Frank was also nearing the end

of his life when the Captain died. Frank's wife Sue had died in 1902. Frank then moved in with his daughter, Jennie Rowland. Lea was taken ill about the time his brother died in February 1904, and was in bad health for about a year before death finally came on February 10, 1905.

Judge Lea's funeral was not as elegant as that of his brother, but it was nevertheless a significant event in Roswell. It was held on Sunday morning, February 12, at 10:00 at the home of Dr. W.T. Joyner, at 206 North Pennsylvania Avenue. The pastors of First Christian Church and First Methodist Episcopal Church, South, Elder C.C. Hill and the Reverend S.R. Twitty, respectively, officiated. The two churches' combined choirs provided the music as the two congregations came together for Judge Lea's funeral as their Sunday morning services.

After the ministers completed their services, the rites were turned over to the Masons for burial at South Side Cemetery. All of the pallbearers were not only Masons but also ex-Confederates, harking back to the days when the judge rode with Quantrill and Shelby.

In announcing the Judge's death, the Daily Record stated:

> Judge Lea was a man of marked individuality, strong in his convictions and opinions; a loyal, law-abiding citizen and a law-enforcing official. His friends were every good citizen in the county .... A more widely-known citizen would be hard to find.

Development of Lea's Block

The center of J.C. Lea's home and business—the 400-block of North Main Street in Roswell—continued to grow and develop both before and after his death. The growth of Lea's home base provides some understanding of the development and growth of the town he so diligently worked to advance.

John W. Poe bought the lots at 400-406 North Main Street from Poe, Lea & Cosgrove Real Estate, deed records indicate, on January 27, 1900. The April 27, 1900, issue of the *Roswell Register* states:

> Last week Mr. Jno. W. Poe had the old adobe store just north of the building occupied [until] eight years ago by Poe, Lea & Cosgrove as a general store, torn down ... to make room for a new store he will soon erect. This old adobe store was one of the original buildings on the town site, before a town was thought of here .... When the building was being torn down ... it was found that the floor joists had sunk into the ground, and the floor was resting on the damp ground and rotting.

It seems probable that the north, west, and south walls of the original Smith-Lea store had been incorporated into the enlarged Poe, Lea, & Cosgrove building in the late 1880s. Further, an on-site inspection of the buildings in 1998 revealed a good possibility that Poe salvaged some of those same walls, using the north wall as a common wall between 400 and his new construction at 402. Poe installed a frame-front, tall building that the August 1900 insurance map shows was occupied by a meat market and a paint store. Interestingly, the original system of house numbering designated the new Poe buildings as 510 and 511 North Main Street, but they were changed by 1905 to 400 and 402. The common adobe wall between 400 and 402 was still in place until the demolition of 1998. Unfortunately, plans to save part of it for an historical marker proved unworkable.

The year after J.C. Lea's death, Poe embarked on another construction project on Lea's home block. Sophie Poe writes in *Buckboard Days* (1936) that her husband built the Pioneer Block (400-406 North Main Street) in 1905, which obviously is incorrect. Available evidence suggests that it was in 1902 that Poe rebuilt 400 and 402 North Main Street and extended them westward to make the building a hundred feet from front to back, built 404 and 406, added a brick common façade on the buildings, and called it the 'Pioneer Block.'

The new commercial units at 408 and 410 North Main Street went up soon after the Pioneer Block. F.J. Peeler acquired the lot at 408 from Ella and Wildy Lea in October 1901, and W.E. Wiseley bought the lot at 410 from the Leas in May 1902. Peeler's location became the New York Store, probably completed in 1902 or early 1903. Wiseley's building was erected in the summer of 1903 and housed Bidwell's candy store.

The first *Roswell City Directory*, 1904, lists E.R. and T.H. Malone at 400 Main, but it doesn't indicate their business. The directory places the *Roswell Daily Record* at 402, and J.H. Croft, jeweler, and J. Bidwell, candy maker, both at 410 (the W.E. Wiseley Building). There is a listing for a boarding house on Main Street, which no doubt was in the former Lea boarding house/residence. The 1905 insurance map shows the Lea house as the 'Clem House' boarding house, so it had reverted somewhat to its original use. Other sources confirm that in 1909 and 1910 the old house was still a boarding house; it was still owned by the Lea heirs.

At undetermined times between 1905 and 1909, Poe put up new buildings in the 100-block of West Fourth, facing Fourth Street between 400 North Main Street and the alley. He used the back wall of 400 as the east wall of the first new addition, which was an office. Later a printing shop was built between the office and the alley.

Additional commercial building projects continued after Lea's death until the entire west side of the 400-block of North Main Street was filled. A long, concrete building at 424 North Main Street was built in 1911 by the Roswell Printing Company and was the home of the *Roswell Register-Tribune* and the *Roswell Morning News*. That same building was later the location of the *Roswell Daily Record* from 1925 until 1971. In late 1997, it became the first on the block to be razed by the city to make room for Pioneer Plaza and city employee parking.

Although the original Roswell store building lost its identity with the construction of the Pioneer Block, the original Smith & Wilburn Hotel at what turned out to be 414-16 North Main Street was still intact for another decade. The building was widely known as Captain Lea's residence. The *Roswell City Directory* for 1912 shows that R.E. Levers had his Roswell Land Company and Levers' Alfalfa Hay Company in the former Lea residence. The lots at 418-20-22 were still vacant in 1912.

All five of the business units between 414 and 522 North Main Street were built as a single project in 1913 called the 'Lea Building.' Wildy Lea owned the lots, and the committee in charge of his affairs put up the building. Erecting the Lea Building required the demolition of Roswell's first building, the Smith & Wilburn Hotel that later served so many years as Captain Lea's house. The approximate location of the building has been preserved in Pioneer Plaza by using different colors of paving tiles to outline the foundation of the house around a Zia symbol.

Lea Cattle Company

While the wrangling over Lea's estate was going on, the Lea Cattle Company finally bit the dust. The principal financier of the company, Horace K. Thurber of New York, apparently died within a year or two after Captain Lea died. Back on July 1, 1885, when the company was forming, 150 bonds of $1,000 each were issued to Thurber. Coupons on the bonds were redeemable semi-annually for six percent interest, and no coupons had been paid since 1891. Thurber's widow, Nancy Thurber, filed suit against the Lea Cattle Company for the value of the bonds and coupons: $232,500. The only officer of the company left in Chaves County was J. Smith Lea, the secretary of the corporation, so he had to deal with the suit.

On February 9, 1906, the Fifth District Court (Civil Docket, Cause Number 854) in Roswell found in Mrs. Thurber's favor and ordered the sheriffs of Chaves and Lincoln Counties to seize "all of the assets" of the cattle company to satisfy the claim. It took some time to figure out just what the assets of the company were, but it was soon determined that the company's assets in Lincoln County were much more extensive than those in Chaves County. Some 185 tracts of forty acres each, totaling more than seven thousand acres, ranged from just east of Fort Stanton, scattered along both sides of the Río Bonito and Río Hondo, mostly south of the latter stream, all the way eastward to the Chaves County line. Lincoln County Sheriff John W. Owen offered the land for public auction at the front steps of the courthouse in Lincoln on August 7, 1906. Others bought small acreages, but the bulk of it went to Nancy Thurber for her high bid of $9,876.

The appraisers determined that in Chaves County there were four tracts of forty acres each, plus some city lots that belonged to Lea Cattle Company. Sheriff K.S. Woodruff offered the lands at auction to the highest bidder on the front steps of the courthouse in Roswell on August 18, 1906. Nancy Thurber bid $5,839 and received title to the property.

By purchasing the assets of the cattle company at the sheriffs' auctions, Mrs. Thurber was able to salvage some of the property her husband had financed more than twenty years earlier. Selling the land put her in a position to recoup a portion of the losses her husband had incurred in his adventures of financing western ranches.

Monuments to Captain Lea

Lea Park. Upon Captain Lea's death, even before his funeral, Editor Harvey Bear of the *Roswell Daily Record* brought up the question of "what will Roswell do to erect a suitable and substantial monument to him who was the pioneer of the Valley?" Bear believed the love for Lea on the part of Roswell residents would prompt the building of some material monument as a fitting tribute to his life and work. Bear thought the appropriate monument would be to complete the Captain's plans for a park and natatorium in downtown Roswell.

> Let this park and natatorium be built as he had planned them by public subscription, and let them stand in the heart of our town christened by the hallowed name he bore—LEA—to be a monument to his life and work, reared on the site of his own homestead, in the town he founded and whose first executive he was, built and dedicated to him by the hands of those who love, honor and cherish his memory.

Apparently, this proposal did not receive enough public enthusaiasm and was never accomplished.

Lea Avenue. About a year after Captain Lea's death, the city council took steps to name a street in his honor. Spring River Avenue, a long, north-south residential street was situated four blocks west of Main Street between Kentucky and Missouri Avenues. Mayor James F. Hinkle introduced a resolution in the February 10, 1905, meeting of the Roswell City Council to change the name of Spring River Avenue to 'Lea Avenue.' It continues to be an important thoroughfare. Carlsbad and other area cities also have streets named for Lea.

Lea Hall. One of the earliest monuments to Captain Lea was the first building of New Mexico Military Institute on North Hill in 1898. The Institute, with which Captain Lea was familiar at the time of his death, was well on its way to becoming the significant power in education that he had always envisioned.

On August 31, 1909, the first Lea Hall burned. The regents acted quickly to replace the building and decided to keep the name 'Lea Hall.' The new structure was dedicated on November 8, 1910, the sixty-ninth anniversary of Lea's birth. The keynote speaker for the ceremony was Thomas C. Lea, Jr., Captain Lea's nephew and father of the noted writer and artist Tom Lea. This later Lea served as mayor of El Paso and was on Pancho Villa's 'hit list.' Captain Lea's son-in-law, Hiram M. Dow, also spoke at the dedication and presented a gift to NMMI of a life-size portrait of Captain Lea on behalf of the Alumni Association. The present whereabouts of the portrait is unknown. Later there was to be a third and a fourth Lea Hall, according to Bill Gibbs.

Two other significant events took place in 1909: NMMI was classified as a Distinguished Military School, an honor which it has retained ever since; and the first section of the Hagerman barracks was built. NMMI became a junior college in 1915. In 1920, it changed from being a Battalion of Infantry to being a Squadron of Cavalry. Numerous facilities were added over the years: *e.g.,* Willson Hall was dedicated in 1928, followed by Cahoon Armory the next year. Pearson Auditorium was built in 1940. In 1948, NMMI became a four-year college until March 1955. The Alumni Memorial Chapel was dedicated in 1975, followed by the L.T. Godfrey Athletic Center in 1980.

In some ways, NMMI came full circle in 1977 when twenty-five young women entered the military school as equal and full-fledged members of the Corps of Cadets, harking back to the first decade of the school's history when girls were enrolled as day students. The Institute served as a role model for other military schools in

NMMI Lea Hall. This is the second Lea Hall, built in 1909. This photograph was taken *c.* 1912-15.
Courtesy: HCSNM, Photo No. 165-E

the nation seeking to serve women on an equal basis with men. Since about 1965, the enrollment at NMMI has stayed around 950 to 1,000 students, of which some twenty percent are women.

Lea County. A permanent and visible tribute to Captain Lea was the creation of a new county named in his honor in 1917. Lea County was formed from the 'Plains Precincts' of eastern Chaves and Eddy counties after a long struggle to win legislative support for the new county.

The first leader of the Lea County movement was Robert Florence Love, founder of Lovington, according to Gil Hinshaw in his 1976 book, *Lea, New Mexico's Last Frontier*. Upon New Mexico's achievement of statehood in 1912, Love represented Eddy County in the new Legislature. He immediately introduced a bill to create 'Heard County' out of the eastern halves of Chaves and Eddy, to be named for Allen C. Heard, a founder of the High Lonesome Ranch and the village of Knowles. The Republican majority in the first Legislature paid little heed to Love's bill, along with about a dozen other similar proposals.

Love asked Paschal S. Eaves, Lovington businessman and rancher, to take up the cause. He was elected to the Legislature in 1916 and introduced a bill to create a new county out of some forty-two townships of eastern Chaves and Eddy Counties, to be called 'Llano County'—Spanish for 'plains.' This ignited a political battle in Chaves County because Progressive Democrats feared the loss of the plains would put the conservative Democrats back into control of Chaves County, and the Republicans supported it because they hoped that would happen.

A petition was sent to the Legislature claiming that ninety-five percent of the residents of the plains were content with the existing counties and that the new county movement was all a plot to make money for Eaves and "his political co-operators." The Chaves County officials wired the Speaker of the House, William Llewellyn, a Republican from Doña Ana County, to oppose the new county proposal. Some taxpayers of Eddy County sent a similar message.

Back in 1889, when Chaves and Eddy Counties were created, Llewellyn was one of Captain Lea's friends in the Legislature who had wanted to name Chaves County after Lea. In 1917, as Speaker, Llewellyn apparently saw a chance to make up for a missed opportunity. In a letter to Ella Lea Dow, written after the close of the legislative session, Speaker Llewellyn told her how he had taken the initiative in changing the name of the proposed county from Llano to Lea. He stated that he argued in the committee hearing that the name 'Llano' would be a problem because people would not know how to pronounce it, never mind how to spell it. Llewellyn claimed that he promised to see that the county bill passed if the sponsors would change the name of the proposed county to 'Lea County.' The agreement was made, the bill was passed, and Governor Washington E. Lindsey signed it on March 7.

Hinshaw observes that most accounts of the history of Lea County have not credited Llewellyn for giving the county its name. Some have attributed the move to Love and Eaves, the recognized founders of Lea County, who may have chosen the Lea name as a ploy designed to win support from Chaves County. Further, George Curry, in his autobiography, takes credit for the name. Former Governor Curry claims that a committee of landowners and businessmen from the area of the proposed new county came to see him in Socorro where he operated a hotel. "I told the delegation that if they would change the name to Lea County, I would help them in the next legislative session."

To the residents of Lea County, J.C. Lea was "... at best a distant hero," according to Hinshaw. With few resources except grass, the new county struggled for many years. Curry claims that wildcatters had already discovered oil and gas around "... the booming new oil town of Hobbs," but in truth it appears that Curry places this petroleum development about ten years ahead of the facts. However, Lea County finally came into its own as one of the richest oil-producing areas in the world.

Lea Statues. When the present Chaves County Courthouse was completed in 1911, the county commissioners desired to put a one-and-a half-times life-size bronze bust of Captain Lea in the rotunda of the new building. The project architects, I.H. & W.M. Rapp of Trinidad, Colorado, and Santa Fe, sent photographs to a

New York Bronze foundry, Jno. Williams, Inc., for an estimate on the bust. The foundry wrote the architects on August 16, 1912, stating the price at $800. Roswell Mayor William Atkinson and many prominent men endorsed the idea and called a public meeting to raise $1,500 for the statue. The effort apparently failed. The bust was never produced and does not appear in any of the itemized accounts presented by the architects to the county commissioners.

Another proposal to erect a statue of Captain Lea was presented by former White Oaks lawyer G.W. Prichard in a letter to the editor of the *Southwestern Dispatch* on December 8, 1925. Prichard wrote that he had been a member of the upper chamber of the Territorial Legislature in 1889:

> ... when the act creating Chaves County was passed. It was to (Lea's) efforts more than to all other influences combined that the bill was passed. It is my belief had it not been for his presence and his wishes, Chaves county would not have been created when it was.
>
> Captain Lea had a statewide acquaintance. All of the old timers knew him, and respected and loved him for his unselfishness, his generosity and his bigness of heart. Those who have survived him would like to see his memory perpetuated in bronze or marble ... . I suggest that an association be organized ... for the purpose of raising funds by private donations with which to erect on the Court House ground at Roswell an appropriate monument, with suitable inscription thereon, to his memory ... .

The Prichard plan attracted some interest but no action.

When the Roswell Museum opened on October 6, 1937, among the exhibits were four busts by sculptor John Raymond Terken. They were of J.C. Lea, 'The Father of Roswell,'; John S. Chisum, 'Cattle King of the Pecos'; J.J. Hagerman, railroad builder and developer; and Amelia Bolton Church, pioneer wife and mother. These four were deemed by the Works Progress Administration's Federal Art Project to have "contributed most to the development and civic improvement of Roswell." The history and art interests of the Roswell Museum eventually parted

Ella Lea Bedell Dow and her second husband,
Lieutenant Governor Hiram M. Dow, with
the bust of her father, Captain J.C. Lea,
by John Raymond Terken, at the
Roswell Museum, 1937.
Courtesy: HCSNM, Photo No. 2133-A

ways, and the Chaves County Historical Society kept the busts. When the Chaves County Historical Museum (the present Historical Center for Southeast New Mexico in Roswell) opened in July 1976, the busts were refinished, placed on new pedestals, and displayed in the museum; they remain on exhibit there.

The Chaves County Historical Society discussed the idea of erecting a statue of Lea, perhaps on his gravesite, in the 1960s. Nothing came of this discussion, however.

As part of the future plans for the Pioneer Plaza on the site of Lea's original home in Roswell, bronze statues are to be placed in appropriate places as funds become available. The centerpiece of the plaza is a large statue of cattle baron John S. Chisum on horseback hazing a Longhorn steer, the result of a campaign headed by Stu Pritchard. The Chisum statue was dedicated with appropriate festivities on March 24, 2001. Plans call for a statue of J.C. Lea to be added to the plaza in the future, according to Pritchard.

Lea Lake. One of the early projects in President Franklin D. Roosevelt's New Deal program to relieve the Great Depression was the first state park in New Mexico—Bottomless Lakes State Park. Cowboys who could not reach the bottom with lariat ropes tied together deemed the deep lakes—sometimes called 'Seven Lakes' by early settlers—'bottomless.' The Federal Emergency Relief Administration undertook the building of a state park with federal funds, completing it in 1934. The main recreational area was established at the largest lake, which was named 'Lea Lake' in honor of Captain J.C. Lea. After more than six decades, Lea Lake continues as the main attraction at Bottomless Lakes.

CONCLUSION

Joseph C. Lea was an interesting and important pioneer and a major developer of Southeast New Mexico. Starting life as a farm boy, he grew up in Tennessee and Missouri as the son of a planter and physician. When the Civil War started, Lea joined the Confederate guerrilla forces of the Missouri-Kansas border conflicts. He served with William Quantrill and Joseph Shelby, but Lea eventually became a captain in charge of his own men.

Lea floundered around the South after the war for a few years. In 1867, he married a widow; she soon died. More instability followed, but in 1875 Lea married Sallie Wildy, the daughter of a Mississippi planter. Lea soon left for New Mexico to try sheep and cattle ranching. After a couple of years in the mountains of northern New Mexico, Lea moved his family and operation to the Pecos Valley. Van Smith had started Roswell earlier, but it was up to Lea to actually develop it. Lea and his second wife Sallie had two children; Sallie died in 1884.

Because of his development of the town, Lea came to be called 'The Father of Roswell.' But Lea was more than just the Father of Roswell. His influence helped keep the Lincoln County War out of Roswell; and Lea played a principal part in securing the services of Pat Garrett as sheriff to rid the country of outlaws, especially Billy the Kid.

Once the violence in Lincoln County subsided, Lea turned his attention to his main priority: big-time cattle ranching. Securing financing from New York businessman Horace Thurber, Lea created the Lea Cattle Company in 1885. The company spread became the largest ranch and had more cattle than any other ranch in Lincoln County and possibly in New Mexico Territory by 1890. The company was organized mainly to raise cattle, but it also had much land for development. To attract settlers to the Pecos

Valley, the company developed both the town and the farmland in the Valley. They built many miles of ditch systems that irrigated thousands of acres of farmland. The company also surveyed additions to Roswell and sold, or sometimes gave, lots to dependable people to establish businesses and homes. The cattle company gradually went under as a result of economic forces beyond Lea's control.

Although Lea was the principal influence in the movement to create Chaves and Eddy counties in 1889, he declined to have a county named after him. Other significant developments in which Lea had a hand were the incorporation of Roswell, the building of the railroad into Roswell, and the growth of the business section.

Goss Military Institute was established in 1891, thanks to J.C. Lea and his third wife Mabel. The school later became New Mexico Military Institute and is one of the premier military schools in the country.

In 1903, the government of Roswell changed from that of a town to that of a city. Lea was drafted to run for—and was elected to serve as—the first mayor until the regular election cycle came up. Sadly, he lived only a few weeks after he took office.

J.C. Lea received many honors and recognitions in his life. His many friends continued to honor his memory after he was gone. Lea experienced his share of sorrow, such as losing two wives to premature deaths, having his only son confined to a mental hospital, and seeing some of his investments fail to produce a return.

J.C. Lea was genuinely a remarkable man who was held in high esteem by everyone who was fortunate enough to come under his influence. Though he started his adult life as a Confederate guerrilla, he went on to become a big-time ranchman in New Mexico Territory and the patriarch of the city of Roswell and the Pecos Valley.

SOURCES CONSULTED

Primary Sources

Ballard Funeral Chapel, Roswell, New Mexico. National Funeral Records: Harry Wildy Lea, Ella Lea Bedell Dow.

Buckner, Barbara; Carlsbad, New Mexico. Personal interview, September 1, 2000.

Chandler, Claudine Sowell. "About Thomas C. Lea." *The Kansas City Genealogist*, Vol. XXII, No. 2.

Chaves County Commissioners' Proceedings. Chaves County Courthouse, Roswell, New Mexico.

Chaves County Deed Records. Chaves County Courthouse, Roswell, New Mexico.

Chaves County Probate Records, No. 7 (Sallie Wildy Lea), No. 113 (J. C. Lea), No. 126 (Frank H. Lea), No. 171 (Mabel Doss Day Lea), No. 1721 (Harry Wildy Lea), No. 4850 (Ella Lea Bedell Dow). Chaves County Courthouse, Roswell, New Mexico.

Chaves County Record of Official Bonds, Book A. Chaves County Courthouse, Roswell, New Mexico.

Church, Amelia Bolton. Papers. Archives, Historical Center for Southeast New Mexico, Roswell.

Colfax County Deed Records. Colfax County Courthouse, Raton, New Mexico.

Crook, Robert. E-mail letter to the author, May 20, 2000.

Dow, Hiram M. Papers. "The New Mexico Military Institute from its Very First Inception." Undated newspaper article, *c.* 1925. Archival Collection No. 760608. Historical Center for Southeast New Mexico, Roswell.

Eckerman, Weldon R.; El Paso, Texas. Letter to the author, January 27, 2000.

Field and Farm, January 5, November 30, 1889.

Fraser, Lea Rohr. Papers. Archival Collection No. 980109, Historical Center for Southeast New Mexico, Roswell.

Garry, Thomas Lea; San Gabriel, California. Telephone interview with the author, November 3, 2000.

George, B.J. *Letters*. Western Manuscript Collection, University of Missouri-Columbia, Columbia, Missouri.

Gibbs, William E.; Roswell, New Mexico. E-mail letter to the author, October 11, 2000.

Gressley, Gene M.; Jacksonville, Oregon. Letters to the author: June 9, 2000; August 30, 2000.

Gravel, Richard H., Master's Thesis, 1954 (Eastern New Mexico University), "The Pecos Valley Railroad, 1889-1906."

Hale, Donald R.; Lee's Summit, Missouri. Letters to the author, February 4, May 7, June 8, July 26, 1999.

James, Jason W. *Memorable Events in the Life of Captain Jason W. James*. Uvalde, Texas: By the Author, 1912.

Kansas City Times, March 10, 1966.

Lea Cattle Company. *Records of Lea Cattle Company, Lincoln County, New Mexico* (Minutes of the Board of Directors, *c*. 1885-1892). In the possession of Mary Herrington, Roswell, New Mexico.

Lea, Tom; El Paso, Texas. Letter to the author, December 29, 1998.

Lincoln County Commissioners Record, Book A, March 1, 1876, to January 9, 1889. Lincoln County Courthouse, Carrizozo, New Mexico.

Lincoln County Deed Records. Lincoln County Courthouse, Carrizozo, New Mexico.

Littell, Max; Roswell, New Mexico. Personal Interviews, Spring 1998.

Long, Lawrence W., III; Yazoo City, Mississippi. E-mail letters to the author, February 28, 2000; March 1, 2, 4, 5, 7, 8, 12, 2000; May 1, 2000; June 30, 2000; August 9, 2000; October 4, 2000.

Miller, J.M. Autobiography. MS in Archives of Historical Center for Southeast New Mexico, Roswell.

Mississippi State Department of Archives and History, Jackson. Wildy, W.W., Capt. Gartley's Company (Yazoo Rangers) Mississippi Cavalry (Confederate). *Company Muster Rolls,* Dec. 10, 1962, to Feb. 28, 1863; May 14, 1862; June 14 to Nov. 1, 1862.

Mora County Deed Records. Mora County Courthouse, Mora, New Mexico.

National Archives, Washington, D. C. James, Jason W., Capt. Farris' Battery (Clark Artillery), Missouri Light Artillery (Confederate). *Company Muster Rolls*, March & April 1862, Sept. & Oct. 1862.

National Archives, Washington, D. C. Lea, T.H., Co. A, 6th Missouri Cavalry (Confederate). *Company Muster Rolls,* Sept. 14. to Dec. 31, 1862.

New Mexico State Records Center and Archives, Santa Fe. *Miscellaneous Corporate Records,* Irrigation documents filed by J.C. Lea, August 15, 1880; September 1, 1880; September 6, 1880.

New Mexico State Records Center and Archives, Santa Fe. *Ritch Collection.* Microfilm Roll 10, Item No. 1900 (J.C. Lea to Gov. Lew Wallace, Dec. 24, 1880), Item No. 1901 (Charles Bowdre to J.C. Lea, Dec. 15, 1880); Roll 25, p. 239 (J.C. Lea to Gov. W.G. Ritch,

October 8, 1880); Roll No. 101, p. 32 ff. (J.C. Lea to Gov. E.G. Ross, July 8, 1885), p. 887 (J.C. Lea to Gov. E.G. Ross, March 29, 1889); Roll No. 103, p. 922 (J.C. Lea to Gov. L.B. Prince, June 11, 1889); Roll No. 104, 453 ff. (J.C. Lea to Gov. L.B. Prince, September 26, 1889); p. 720 (J.C. Lea to Gov. L.B. Prince, November 15, 1889); Roll 105, p. 814 (J.C. Lea to Gov. L.B. Prince, March 18, 1890); Roll 109, p. 376 (J.C. Lea to Gov. L.B. Prince, July 14, 1891); Roll No. 135, p. 160 (G.A. Richardson to Gov. M.A. Otero).

New Mexico State Records Center and Archives, Santa Fe. *Territorial Archives of New Mexico*, Microfilm Roll No. 99, p. 126 (J.C. Lea to Gov. W.G. Ritch, (n.d.).

New Mexico State Records Center and Archives, Santa Fe, New Mexico Territorial House of Representatives. "An Act to Create the Counties of Chaves and Eddy," February 25, 1889.

Nelson, Morgan. Personal Interview, August 1, 1998.

North Spring River Centre Ditch Company, Roswell, New Mexico. Articles of Incorporation, October 1, 1881. Archives, Historical Center for Southeast New Mexico, Roswell.

Pecos Valley Artesian Conservancy District, Roswell, New Mexico. Archives.

Pecos Valley Register, 1888-1890.

Redfield, Georgia B. Interviews with Old-Timers, Federal Writers' Project, 1936-1940: "C. D. Bonney," Item 33. "Patrick H. Boone," Item 1. "Ella Lea Dow," Item 77.

________. "Captain Joseph Callaway Lea." MS, 1938. Archives, Historical Center for Southeast New Mexico, Roswell.

Reed, Verna; Carlsbad, New Mexico. E-mail letters to the author, March 2, 3, 6, 8, 9, 10, 11, 14, 22, 23; April 3; August 1, 2, 3; September 21, 2000.

________. Personal interview, September 1, 2000.

Roswell City Council. *Ordinances*. Roswell City Hall, Roswell.

Roswell City Council. *Proceedings*. Roswell City Hall, Roswell.

Roswell City Council. *Resolutions*. Roswell City Hall, Roswell.

Roswell, City of. *South Park Cemetery Records, 1883-2001.* Archives, Historical Center for Southeast New Mexico, Roswell.

Roswell Daily Record.

Roswell Evening News.

Roswell Morning Dispatch.

Roswell Record.

Roswell Register.

Roswell Register-Tribune.

Southwestern Dispatch.
Taos County Deed Records. Taos County Courthouse, Taos, New Mexico.
U.S. Army. *War of the Rebellion, Official Records of the Union and Confederate Armies, Series I,* Vol. XLI, Part IV (1893), pp. 128-29, 675, 901; 951-52.
U.S. Bureau of the Census. *Eighth Census of the United States, 1860.* Jackson County, Missouri, p. 348.
U.S. Bureau of the Census. *Eighth Census of the United States, 1860.* Yazoo County, Mississippi, p. 977.
U.S. Bureau of the Census. *Ninth Census of the United States, 1870.* Yazoo County, Mississippi, p. 362.
U.S. Bureau of the Census. *Seventh Census of the United States, 1850.* Yazoo County, Mississippi, p. 485
U.S. Bureau of the Census. *Sixth Census of the United States, 1840.* Yazoo County, Mississippi, Abstract.
Torrez, Robert J.; New Mexico State Historian, Santa Fe. E-mail letter to the author, October 13, 2000.
White, Jim; Farmington, New Mexico. Personal Interview, August 22, 1998.
Younger, Cole. *The Story of Cole Younger, by Himself.* Provo, Utah: Triton Press, 1988 (reprint of 1903 edition).

Secondary Sources

Bonney, Cecil. *Looking Over My Shoulder: Seventy-five Years in the Pecos Valley.*Roswell: By the Author, 1971.
Breihan, Carl W. *Quantrill and his Civil War Guerrillas.* Denver: Sage Books, 1959.
Burch, John P. *Charles W. Quantrell.* N.p., n.d.
Coan, Charles F. "Joseph C. Lea," *A History of New Mexico*, Volume I. Chicago: The American Historical Society, 1925.
Curry, George. *George Curry 1861-1947: An Autobiography.* Albuquerque: University of New Mexico Press, 1958.
D'Abadie, Bernice. "History of Early Ditches of Chaves County." Address to Chaves County Historical Society, Sept. 11, 1989. Audiotape, Archives, Historical Center for Southeast New Mexico, Roswell.
DeCell, Harriet, and Prichard, JoAnne. *YAZOO: Its Legends and Legacies*. Yazoo City, Mississippi: Yazoo Delta Press, 1976.

Dills, Lucius. "History of Chaves County." Unpublished manuscript, *c.* 1943, 300 pp. Archives, Historical Center for Southeast New Mexico, Roswell.

________. *ROSWELL: Some Facts and Observations Relative to Its Settlement and Early Growth.* Roswell: Chaves County Archaeological and Historical Society, 1933.

Eakin, Joanne Chiles. "Doctor Pleasant Lea," *Kansas City Genealogist*, Vol. 34, No. 3.

________, and Hale, Donald. R. *Branded as Rebels.* Independence, Missouri: Wee Print, 1993.

Edwards, John N. *Shelby and His Men: or, The War in the West.* Cincinnati: Miami Printing and Publishing Co., 1867.

Fellman, Michael. *Inside War: The Guerrilla Conflict in Missouri During the American Civil War.* New York: Oxford University Press, 1989.

Fleming, Elvis E. "A Brief History of the 400 Block of North Main, 1866-1913," *Vision Magazine, Roswell Daily Record,* April 3, May 1, June 5, August 7, 1998.

________. "A Brief History of the Grazing Industry in Chaves County, New Mexico." *Roswell Daily Record*, January 1997.

________. "Blashek's Mill," *Treasures of History II: Chaves County Vignettes.* Roswell: Chaves County Historical Society, 1991.

________. *J.B. 'Billy' Mathews: Biography of a Lincoln County Deputy.* Las Cruces: Yucca Tree Press, 1999.

________. "Joshua P. Church, Early Roswell Entrepreneur," *Southern New Mexico Historical Review,* January 2000.

________. "Creation of Chaves County," *Treasures of History II: Chaves County Vignettes.* Roswell: Chaves County Historical Society, 1991.

________. "Jaffa, Prager & Company, Roswell's First Department Store," Treasures *of History III: Southeast New Mexico People, Places, and Events.* Roswell: Historical Society for Southeast New Mexico, 1995.

________. "J.J. Hagerman and the Pecos River Railroad," *The Permian Historical Annual*, 1978.

________. "Judge Frank H. Lea Rode with Quantrill's Raiders," *Vision Magazine, Roswell Daily Record,* May 5, 2000.

________. "Lucius Dills," *Treasures of History II: Chaves County Vignettes.* Roswell: Chaves County Historical Society, 1991.

________. "Milo Pierce, Lincoln County War Veteran," *Treasures of History III: Southeast New Mexico People, Places, and Events.* Roswell: Historical Society for Southeast New Mexico, 1995.

________. "Nath Hendricks was Tried Twice for Murdering Deputy Will Rainbolt," *Vision Magazine, Roswell Daily Record,* August 4, 2000.

________. "Roswell's First Train Arrived in 1894," *Treasures of History III: Southeast New Mexico People, Places, and Events.* Roswell: Historical Society for Southeast New Mexico, 1995.

________. "'Sockless' Jerry Simpson: The New Mexico Years, 1902-1905," *New Mexico Historical Review*, January 1994.

________. "Stagecoach Lines of the Pecos Valley," *Treasures of History III: Southeast New Mexico People, Places, and Events.* Roswell: Historical Society for Southeast New Mexico, 1995.

________. "Van C. Smith, Founder of Roswell," *Vision Magazine, Roswell Daily Record,* September 4, October 2, 1998.

________, and Huffman, Minor S. (eds.). *Roundup on the Pecos.* Roswell: Chaves County Historical Society, 1978.

________, and Williams, Ernestine Chesser. *Treasures of History II: Chaves County Vignettes.* Roswell: Chaves County Historical Society, 1991.

________, and Williams, Ernestine Chesser. *Treasures of History III: Southeast New Mexico People, Places, and Events.* Roswell: Historical Society for Southeast New Mexico, 1995.

Fulton, Maurice Garland. *Roswell In Its Early Years.* Roswell: Privately printed, 1963.

Gaddy, Jerry J. *Dust to Dust: Obituaries of the Gunfighters.* Fort Collins, Colorado: Presidio Press, 1977.

Garrett, Pat F. *The Authentic Life of Billy, the Kid.* Edited by Maurice Garland Fulton. New York: The Macmillan Co., 1927.

Gerling, George F. "Lea, Wildy, and Dow Families," *Roundup on the Pecos*. Roswell: Chaves County Historical Society, 1978.

George, Todd M. *Just Memories and Twelve Years with Cole Younger.* N. p., 1959.

Gibbs, William E. and Jackman, Eugene T. *New Mexico Military Institute: A Centennial History.* Roswell: New Mexico Military Institute Centennial Commission, 1991.

Gressley, Gene M. *Bankers and Cattlemen.* New York: Alfred A. Knopf, 1966.

Hail, Marshall. "San Augustine Ranch House," *Frontier Times*, August-September 1969.

Hale, Donald R. *Civil War and Reconstruction in Independence, Missouri.* N. p., n.d.

________. *Jackson County and the Civil War.* Independence, Missouri: Two Trails Publishing, n.d.

________. *The History of Lee's Summit.* Lee's Summit, Missouri: Lee's Summit Chamber of Commerce, 1993.

Hart, Robert L. "Ozanne and Company Stage Line of White Oaks," *Southern New Mexico Historical Review*, January, 1996.

Hickman, W. Z. *History of Jackson County, Missouri.* Topeka: Historical Publishing Co., 1920.

Hinshaw, Gil. *Lea, New Mexico's Last Frontier.* Hobbs: Hobbs News-Sun, 1976.

Jacobsen, Joel. *Such Men as Billy the Kid: The Lincoln County War Reconsidered.* Lincoln, Nebraska: University of Nebraska Press, 1994.

"Joseph C. Lea," *Illustrated History of New Mexico.* Chicago: The Lewis Publishing Co., 1895.

Josephy, Alvin M., Jr. *The Civil War in the American West.* Knopf, 1991.

Keleher, William A. *The Fabulous Frontier.* Albuquerque: University of New Mexico Press, 1962.

________. *Violence in Lincoln County 1869-1881.* Albuquerque: University of New Mexico Press, 1957.

Kelly, J. R. *A History of New Mexico Military Institute, 1891-1941.* Albuquerque: University of New Mexico Press, 1953.

Larson, Carole. *Forgotten Frontier: The Story of Southeastern New Mexico.* Albuquerque: University of New Mexico Press, 1993.

Lankford, Rose Mary. *The Encyclopedia of Quantrill's Guerrillas.* Evening Shade, Arkansas: By the Author, 2001.

Lipsey, John J. *The Lives of James John Hagerman.* Denver: Golden Bell Press, 1968.

McCorkle, John. *Three Years with Quantrell: A True Story.* New York: Buffalo-Head Press, 1966.

Monaghan, Jay. *Civil War on the Western Border, 1854-1865.* Little Brown & Co., 1955.

Moore, John C. *Confederate Military History, Volume IX: Missouri.*

New Mexico Military Institute, *The Bronco.* Roswell: New Mexico Military Institute, 1902, 1903, 1918.

Nolan, Frederick. *The Lincoln County War: A Documentary History.* Norman: University of Oklahoma Press, 1992.

________. "Van C. Smith: 'A Very Companionable Gentleman," *New Mexico Historical Review*, April 1997.

________. *Pat F. Garrett's 'The Authentic Life of Billy, the Kid'* Norman: University of Oklahoma Press, 2000.

________. *The West of Billy the Kid.* Norman: University of Oklahoma Press, 1998.

Nunn, W.C. *Escape From Reconstruction.* Fort Worth: Texas Christian University, 1956.

O'Flaherty, Daniel. *General Jo Shelby: Undefeated Rebel.* Chapel Hill: University of North Carolina Press, 1953.

Padgitt, James T., "Captain Joseph C. Lea, the Father of Roswell," *West Texas Historical Association Year Book* (1959).

________, "Early History of New Mexico Military Institute," *West Texas Historical Association Year Book* (1958).

Peters, Ronald Luther. "The New Mexico Military Institute: Development of Its Campus Plan and Architectural Style from 1890-1930." Thesis, Arizona State University, 2001.

Poe, Sophie A. *Buckboard Days.* Caldwell, Idaho: The Caxton Printers, 1936.

Redfield, Georgia A. "They Came Riding Across the Plains of New Mexico," *The Cattleman*, October 1946.

Rhodes, Eugene Manlove. *Pasó por Aquí.* Boston: Houghton Mifflin Co., 1949.

Roswell Lodge, No. 18, Ancient, Free and Accepted Masons. Temple Book. Roswell: Roswell Lodge No. 18, 1914.

Shinkle, James D. *Fifty Years of Roswell History, 1867-1917.* Roswell: By the Author, 1964.

________. *Reminiscences of Roswell Pioneers.* Roswell: By the Author, 1966.

________. *Seventy-Seven Years of Roswell School History, 1881-1958.* Roswell: By the Author, 1958.

Sons of Confederate Veterans, "Maj. Gen. Joseph Orville Shelby." Sedalia: Missouri Division, Internet Website, 1998.

Thrapp, Dan L. "Lea, Joseph Calloway," *Encyclopedia of Frontier Biography.* Spokane: Arthur H. Clarke Co., 1994.

Tuska, Jon. *Billy the Kid: His Life and Legend.* Albuquerque: University of New Mexico Press, 1994.

Utley, Robert M. *Billy the Kid: A Short and Violent Life.* Lincoln, Nebraska: University of Nebraska Press, 1989.

Westphall, Victor. "The Public Domain in New Mexico, 1854-1891," *New Mexico Historical Review,* January 1958.

White Sands Missile Range, Internet Website, 2000.

William Clarke Quantrill Society, *The No Quarterly*, January 1999.

Williams, Ernestine Chesser. "Captain Jason W. James," *Treasures of History II: Chaves County Vignettes.* Roswell: Chaves County Historical Society, 1991.

________. "Organization of Chaves County," *Treasures of History II: Chaves County Vignettes.* Roswell: Chaves County Historical Society, 1991.

Williams, Jerry E., and McAllister, Paul E. (eds.). *New Mexico in Maps.* Albuquerque: University of New Mexico Press, 1979.
Wilson, John P. *Merchants, Guns & Money: The Story of Lincoln County and Its Wars.* Santa Fe: Museum of New Mexico Press, 1987.
Zornow, William F. *KANSAS: A History of the Jayhawk State.* Norman: University of Oklahoma Press, 1957.

INDEX

LINEAGE OF JOSEPH C. LEA AND SALLIE WILDY LEA

JOSEPH C. LEA
B: Nov. 8, 1841
P: Cleveland, TN
M: Feb. 3, 1875
D: Feb. 4, 1904
P: Roswell, NM

Pleasant J. G. LEA
B: Nov. 6, 1807
P: Cleveland, TN
M: 1838
D: 1862
P: Jackson Co., MO

Major LEA, Jr.
B: Feb. 26, 1775
P: Tennessee
M: c. 1800
D: Aug. 8, 1821
P: Cleveland, TN

Major LEA
B: 1741
P: North Carolina
D: Nov. 19, 1843
P: Powell's Valley, TN

James LEA
B: 1718
P: Cheshire, England
D: June 2, 1788
P: North Carolina

Anne TOLBERT
B:
P:
D:
P: North Carolina

Elizabeth HERNDON
B:
D:
P: Tennessee

Rhoda JARNAGIN
B: Jan. 26, 1778
P: Jefferson Co., TN
M: c. 1800
D: Dec. 21, 1852
P: Cleveland, TN

Thomas JARNAGIN
B:

Mary WITT
B:

Lucinda CALLAWAY
B: 1821
D: 1857
P: Jackson Co., MO

Joseph CALLAWAY
B:
D:

Nancy (?)
B:
D:

Joseph WILDY
B: July 22, 1733
P: Northumberland Co.,VA

Joseph Wildy

Elizabeth C. Cottrell
D: 1765